D0305237

AFTER THE FLOOD

Also by John Nichol

NON-FICTION
The Red Line
Arnhem
Medic
Home Run
Tail-End Charlies
The Last Escape

AUTOBIOGRAPHY
Tornado Down
Team Tornado

FICTION
Point of Impact
Vanishing Point
Exclusion Zone
Stinger
Decisive Measures

JOHN NICHOL

AFTER THE FLOOD

WHAT THE DAMBUSTERS DID NEXT

WILLIAM
COLLINS

William Collins
An imprint of HarperCollins*Publishers*
1 London Bridge Street
London SE1 9GF
www.WilliamCollinsBooks.com

First published in Great Britain by William Collins in 2015

1 3 5 7 9 10 8 6 4 2

A catalogue record for this book is
available from the British Library.

HB ISBN 978-0-00-810031-5
TPB ISBN 978-0-00-810084-1

Printed and bound in Great Britain by
Clays Ltd, St Ives plc

MIX
Paper from
responsible sources
FSC
www.fsc.org
FSC® C007454

For Sophie

*This book is dedicated to all the members of 617 Squadron,
on the ground and in the air, who served during the
Second World War with such courage and fortitude.
Their sacrifice will never be forgotten.*

Contents

Acknowledgements

Many people willingly gave their valuable time and considerable expertise whilst I researched and wrote this book. Without their input, my task would have been almost impossible. It is difficult to mention every person individually but I am eternally grateful to them all. My sincere thanks also go to:

Rob Owen, the 617 Squadron official historian, who introduced me to many of the veterans, provided countless personal accounts and pictures, and, most importantly, read and corrected the draft manuscript. Rob's input was invaluable.

Charles Thompson and the team from TVT Productions (especially Willard, Matt, Thea and Victoria) – it was Charles who invited me to help make and present the TV documentary *What the Dambusters Did Next*, which was the inspiration for this book.

Chris Ward, the author of *Dambusters: The Forging of a Legend*, which was a unique reference source for 617 Squadron's day-to-day operations. Chris also helped me locate a number of the veterans' surviving relatives.

Charles Foster, author of *Breaking the Dams*, who accompanied me on a trip to the Dortmund–Ems canal and provided a number of deceased veterans' personal accounts.

Gordon Leith and Peter Elliot at the RAF Museum Hendon who guided me through their archives.

My editor Arabella Pike and the truly excellent team at William Collins for their expertise and patience. And Neil Hanson for his dedication to the cause.

My friend and agent Mark Lucas, who has always been there to offer guidance, advice and encouragement.

I am indebted to countless other historians, authors and researchers who offered invaluable advice and sources. It is impossible to name them all, but the following provided important leads, accounts and background information: Tom Allett, Alex Bateman, Malcolm Brooke, Werner Bühner, Peter Carlyle-Gordge, Jan Cheney, Axel Frick, Derek Gill, John Gumbley, Bruce Hebbard, Martin Mace, Dr Marcus Meyer, Richard Morris, Valérie Noël, Joe O'Connor, Alan Parr, Mark Postlethwaite, Peter Rice, John Saunders, Mary Stopes-Roe, Bruce Vigar and Terry Wiltshire.

To my wife Suzannah and daughter Sophie for their ever-present love and support.

Finally, to the countess veterans and their relatives who told me their stories, some long supressed, I am truly grateful – I could use only a fraction of the incredible accounts I heard, but I hope I have done you all justice.

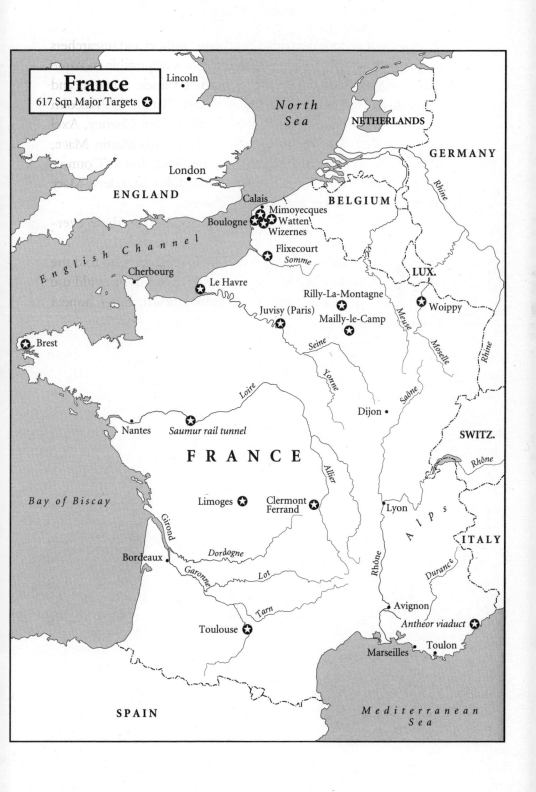

France
617 Sqn Major Targets ⭐

Lincoln
North Sea
NETHERLANDS
GERMANY
London
ENGLAND
Calais
Mimoyecques
BELGIUM
Boulogne ⭐⭐⭐ Watten
Wizernes
Flixecourt
Somme
LUX.
English Channel
Cherbourg
Le Havre
Rilly-La-Montagne
Woippy ⭐
Juvisy (Paris) ⭐
Mailly-le-Camp ⭐
Seine
Brest ⭐
Yonne
Meuse
Moselle
Rhine
Loire
Dijon •
Saône
SWITZ.
Nantes •
Saumur rail tunnel ⭐
FRANCE
Rhône
Bay of Biscay
Limoges ⭐
Clermont Ferrand ⭐
Allier
Lyon •
Alps
ITALY
Bordeaux •
Dordogne
Girond
Garonne
Lot
Rhône
Durance
Rhône
Tarn
Avignon •
Antheor viaduct ⭐
Toulouse ⭐
Toulon •
Marseilles •
SPAIN
Mediterranean Sea

Germany
617 Sqn Major Targets ✪

0   25   50   75   100
miles

Kiel

Rostock

Wilhelmshaven

Bremerhaven

Hamburg

Politz oil refinery
Stettin

Bremen
Arbergen
railway bridge

Aller

Elbe

Oder

NETH.

Dortmund–Ems
Canal

Osnabrück

Hannover

Ehmen

Berlin

Potsdam

Waal

Münster

Ems

Bielefeld

Brunswick

Lübben

Eindhoven

Essen

Ruhr

Möhne dam

Kassel

Düsseldorf

Eder dam

Sorpe dam

Leipzig

Mönchengladbach

Cologne

Bonn

Weimar

Dresden

Rhine

G E R M A N Y

BELG.

Koblenz

Moselle

Frankfurt

Main

LUX.

Bayreuth

Prague

Metz

Nuremberg

CZECH.

Karlsruhe

FRANCE

Stuttgart

Danube

Strasbourg

Danube

Rhine

Kembs barrage

Munich

Linz

Berchtesgaden

AUSTRIA

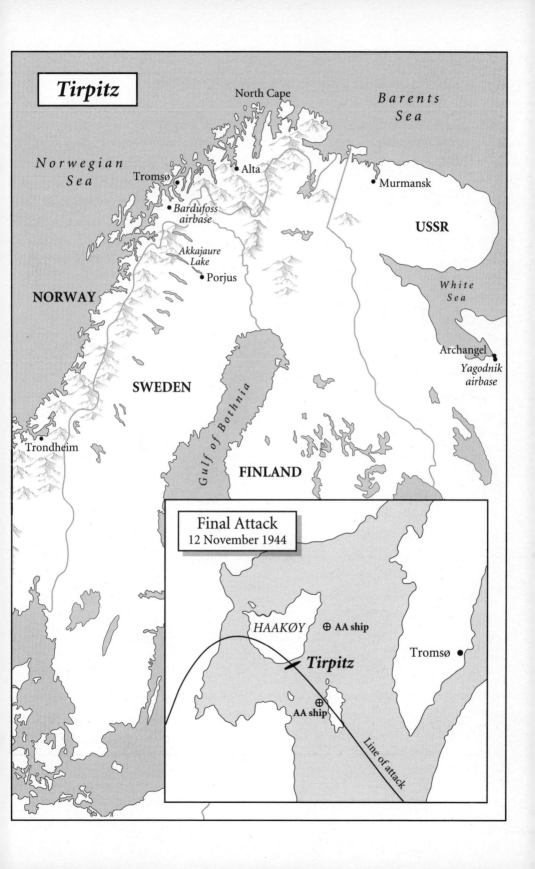

Tirpitz

North Cape

*Barents
Sea*

*Norwegian
Sea*

• Alta

Tromsø •

• Murmansk

*Bardufoss
airbase*

USSR

*Akkajaure
Lake*

NORWAY

• Porjus

*White
Sea*

Archangel •
*Yagodnik
airbase*

SWEDEN

FINLAND

Gulf of Bothnia

• Trondheim

Final Attack
12 November 1944

HAAKØY

⊕ AA ship

Tirpitz

Tromsø •

⊕
AA ship

Line of attack

Author's Note

'*Après moi, le déluge ...*'

King Louis XV's last words became the motto of the most famous bomber squadron in history – 617 Squadron RAF – the Dambusters. Their role in Operation Chastise – the attack on the Möhne, Eder and Sorpe dams at the heart of the industrial Ruhr valley on 17 May 1943 – has been celebrated in print and on screen for more than seventy years. But what 617 Squadron did in the aftermath of this iconic raid is far less well known.

617 Squadron was a specialist squadron, formed from some of the RAF's most brilliant and experienced aircrews for one specific task: breaching the dams. However, British commanders were soon finding other targets for their elite Dambusting squadron, and it was to play the lead role in a series of much less well-known but almost equally eye-catching attacks that destroyed some of the Nazis' most deadly weapons and wrecked key parts of Germany's industrial infrastructure.

617 Squadron's devastating raids caught the imagination, raised the morale of the British public and made headlines around the world. More important, they also helped to tip the balance of hostilities in the Allies' favour, saved countless thousands of lives and arguably contributed to shortening the war.

CHAPTER I

The Dams

A 617 Squadron Lancaster takes off for the iconic raid on the German dams

During the dark days of 1940, 1941 and the early part of 1942, when the British public had been forced to swallow an unremitting diet of blood, sweat, tears, toil and gloom, the RAF and Bomber Command had offered almost the only glimmers of hope. Despite the propaganda spin, while the evacuation of Dunkirk reduced the scale of disaster, it was a disaster nonetheless. British defeats by the Afrika Korps on the battlefields of North Africa and by the Japanese in the Far East – where the surrender at Singapore on 15 February 1942 was not just the greatest humiliation in Britain's military history, but the moment from which the end of Empire could be said to have begun – continued the string of military reverses. Only in the air, where RAF Spitfires and Hurricanes had defeated the German bombers in the Battle of Britain and Bomber Command was relentlessly taking the war to Germany's industrial cities, could Britain be said to be on the offensive.

In late 1942, the victory at the Second Battle of El Alamein ending on 3 November, and the lifting of the Siege of Malta on 11 December, coupled with new tactics at sea which were reducing, though not eliminating, the U-boat menace, suggested the military tide might finally be beginning to turn, but Britain's commanders and people still looked to the air for proof that Germany was paying the price for its aggression.

A four-month bombing campaign against the cities of the Ruhr, aiming to pulverise and paralyse the heavy industries based there and so to disrupt Germany's war production, had begun on 5 February 1943 with the bombing of Essen. But arguably, the most crucial targets were the string of dams in the hills flanking the Ruhr. They not only generated some of the power the heavy industries required, but also supplied drinking water to the population, pure water for steel-making and other industrial processes, and the water that fed the canal system on which the Ruhr depended, both to move raw materials to the factories and to carry finished products – aircraft parts, tanks, guns and munitions – away from them.

However, attempts by Main Force – as the squadrons carrying out mass bombing raids in Bomber Command were known – to attack small targets with the required accuracy had so far proved ineffective. Where the requirement was for saturation bombing over a broad area, Main Force could be brutally effective, as the thousand-bomber raid that devastated Cologne at the end of May 1942 had already demonstrated. But regular success in bombing individual targets – particularly if they were as difficult to access and as ferociously defended as the dams – had proved elusive.

Attacks by torpedo bombers like the Bristol Beaufort and the Fairey Swordfish were foiled by a lack of suitable weapons and by heavy steel anti-torpedo nets strung across the waters of the dams to protect them. A more radical solution was needed and the British engineering genius Barnes Wallis supplied it: the Upkeep 'bouncing bomb', a cylindrical weapon like a heavyweight depth charge, imparted with backspin to rotate it at 500 revs a minute. If dropped at the right height and distance from the dam, the bomb would skim like a pebble thrown across the surface of the water, bounce over the top of the torpedo nets and strike the dam wall. The backspin would then hold it against the face of the dam as it sank below the water before detonating to blow, it was hoped, the dam apart.

Having demonstrated the theoretical effectiveness of the bomb, all that was needed was a squadron of bomber crews capable of delivering it with sufficient accuracy. Since no such squadron existed, it became necessary to create one – 617 Squadron – and at the end of March 1943 recruitment of suitably skilled and experienced crews began under the leadership of Wing Commander Guy Penrose Gibson. He was a man with a glowing reputation as a fearless pilot, willing to take off in even the most marginal weather and attack the most heavily defended targets, whether capital ships or military or economic targets. The head of Bomber Command, Air Marshal Arthur 'Bomber' Harris, later described Gibson as 'the most full-out fighting pilot' under his command and 'as great a warrior as this island ever bred'.

Gibson had flown two full tours on bombers and one on night-fighters, completing the astonishing total of 172 sorties even before joining the new 617 Squadron. As a Squadron Leader and then a Wing Commander, he was as ruthless in screening his crews as he was aggressive in facing the enemy. That ruthlessness and an often abrasive and patrician manner, particularly with NCOs and 'other ranks', made him enemies – some of his ground crews nicknamed him 'The Boy Emperor' – but none could deny his courage or skill as a pilot, which were reflected by the Distinguished Flying Cross and Bar and Distinguished Service Order and Bar he had already been awarded prior to joining 617 Squadron.

Gibson was given a free hand in choosing men – all volunteers – from among crews who had already completed, or nearly completed, two tours of operations. The screening process continued even after they were chosen; Gibson posted two crews away from the squadron after deciding they were not up to the task and a third crew chose to leave after their navigator was also deemed unsatisfactory by Gibson. Intensive training of the remainder lasted for several weeks. In April 1943 alone, his crews completed over 1,000 flying hours, and at the end of that time Gibson reported that they could fly from pinpoint to pinpoint at low level in total darkness, fly over water at an altitude as low as 60 feet and carry out precision bombing with remarkable accuracy. They practised the raid itself using reservoirs in the Peak District and Rutland as substitutes for the Ruhr dams, and after a full-scale dress rehearsal on 14 May 1943, simulating the routes, topography and targets of the actual raid, Gibson pronounced it 'completely successful'. The Dambusters – though no one yet called them that – were ready to take flight.

They had been training for six weeks in the utmost secrecy for a low-level bombing mission, but none of them knew the actual targets until the final briefing on the day the raid was to be launched. When they were told, every man in the room felt a stab of fear, 'and if they said any different,' says air gunner Fred Sutherland, 'they'd be lying, because it looked like a real suicide run.'[1]

They were to target three dams – the Möhne, Eder and Sorpe – and their destruction would cause catastrophic flooding in the Ruhr valley and massive disruption to power generation, water supply, canal transport, agriculture, coal-mining, steel-making and arms manufacturing. If the raid successful.

★　　★　　★

On the nights she was not on duty, telephone operator Gwyn Johnson would often lie in bed, waiting in vain for sleep to come. She could sometimes hear the low, grumbling engine note of the bombers of 617 Squadron on their nearby base breaking the stillness of the night as, one by one, each pilot fired up his Lancaster's four Rolls-Royce Merlin engines and began to taxi from the dispersal areas to the end of the runway. She could picture them waiting until a green light flashed from the runway controller's caravan, when each pilot in turn pushed the throttles to the stops and his heavy-laden bomber began rumbling slowly down the runway. The noise swelled as one by one they lumbered into the air before finally disappearing into the night, the roar of the engines fading again to a sound like the distant rumble of thunder from a summer storm.

Not knowing whether her husband, Bomb Aimer Sergeant George 'Johnny' Johnson, was flying with them on any particular night, she often fell into an uneasy, fitful sleep, and a few hours later, as dawn approached, would wake again as she heard the aircraft return, no longer in a compact formation, but spread across the sky.

Hours later – if he had time – she might meet her husband for a snatched cup of tea in a cafe. He'd tell her if he'd been flying the night before and, though he gave her only the sketchiest details of the mission he'd flown, the look on his face was enough to tell her whether his comrades had returned safely or, more often than not, when one or more aircraft would have failed to return.

Yet, says Johnson, recalling those events seventy years later, 'Gwyn had the same confidence as me that I would always come back. We both had an absolute belief we would survive.'[2] Johnson, a country boy from a small Lincolnshire village with a habit of talking out of the side of his mouth as if all his conversation was 'hush-hush' and on a 'need to know' basis, was the bomb-aimer in American pilot Joe McCarthy's crew.

Johnson and the rest of Joe McCarthy's crew had all finished their thirty-op tour of duty with their previous squadron, 97, two months before the Dams raid, when they took part in a mass raid on the docks at St-Nazaire. 'I could have called it a day after my thirty ops,' Johnson says, 'but I wanted to go on.' After completing a tour, it was standard practice to be given a week's leave before returning to duty, and Johnson and his fiancée, Gwyn, had arranged to get married on 3 April 1943, during his leave. They'd met when he was at the aircrew receiving centre at Babbacombe in 1941. He was walking down the street with his mate when they saw two young ladies walking towards them, and Johnson recalls coming up with 'the corniest chat-up line. I said, "Are you going our way?" to which one of them replied, "That depends on which your way is!" That was Gwyn, that was the start and I never looked back!'

However, the week before the wedding, McCarthy, now the proud holder of a DFC, called the crew together and told them that Wing Commander Guy Gibson had asked him to join a special squadron being formed for a single operation. His crew at once volunteered to go with him for that one trip. The first thing most of the aircrew noticed when they joined 617 Squadron was how experienced most of their fellows were – 'lots of gongs on chests' as Johnny Johnson recalls. 'With all this experience, we were obviously up for something special.' But he had 'no idea what 617 Squadron was going to do,' he says, 'not a clue,' but when he wrote to Gwyn and told her he was going on another op, he added, 'Don't worry, it's just one trip. I'll be there for the wedding.'

Her reply was brief and to the point: 'If you're not there on April the third, don't bother.'

He thought that it would still be fine until they arrived at the new squadron's base, Scampton, where one of the first things they were told was that there would be no leave for anyone until after the operation. However, McCarthy then marched his entire crew into Guy Gibson's office and told him, 'We've just finished our first tour; we're entitled to a week's leave. My bomb-aimer's supposed to be getting married on the third of April and he's gonna get married on the third of April.'

The patrician Guy Gibson didn't like being told what to do by anyone, least of all an NCO, and an American to boot, but in the end he granted them four days' leave and Johnson's wedding went ahead. Wives of married aircrew weren't allowed to live on base, but Gwyn, a military telephone operator, was posted to Hemswell, eight miles from Scampton, and she and Johnny could meet in Lincoln on their days or evenings off. The last buses for Hemswell and Scampton both left Lincoln at nine o'clock, and Johnson invariably ended up on the Hemswell bus and then had to walk the eight miles back to Scampton. 'I was one of the few who had a wife close by,' he says, 'and it was important for me to get away from the squadron atmosphere to "normal" married life.

We never talked about the war and certainly not about any fears or concerns for the future. Our time together was time away from the war.'[3]

Johnny and Gwyn Johnson on their wedding day

*   *   *

Johnny Johnson and the rest of Joe McCarthy's crew were one of nineteen Lancaster crews from 617 Squadron, a total of 133 airmen, who set out for Germany on that 'one special op' on the evening of 16 May 1943. At 9.28 that evening, a few minutes before the sun set over the airfield, the runway controller at RAF Scampton flashed a green Aldis lamp. The four Merlin engines of the Lancaster bomber, already lined up on the grass runway, roared up to maximum power and the aircraft thundered into the sky. 617 Squadron was on its way to Germany, and though none of the participants knew it at the time, into the history books.

Two of the nineteen crews turned back even before reaching the enemy coast; one accidentally losing their bomb over the

North Sea, another with a flak-damaged radio. Another Lancaster was lost altogether after its pilot, Flight Lieutenant Bill Astell, who like his peers was flying at extreme low level, flew straight into an electricity pylon, killing himself and his entire crew. Sixteen aircraft now remained.

617 Squadron's leader, Wing Commander Guy Gibson, led the remainder of the first wave to their first target, the Möhne dam. Each crew would have to drop their bomb with total precision, flying at a speed of 232 mph, exactly 60 feet above the surface of the dam. There was no margin of error – dropped from too low an altitude, any water splash might damage the aircraft; too high and the bomb might simply sink or shatter. If dropped too far away, it could sink before reaching the dam; too close, and it might bounce over the parapet.

While the remainder circled at the eastern end of the lake, Gibson began his bomb-run, swooping down over the forested hillside and roaring across the water towards the dam, its twin towers deeper black outlines against the darkness of the night sky. The Upkeep bomb was already spinning at 500 rpm, sending juddering vibrations through the whole aircraft – 'like driving on a cobbled road' as one crewman described it.[4] Gibson had just fourteen seconds to adjust his height, track and speed before the moment of bomb-release. Alerted by the thunder of the Lancaster's engines, the German gun-crews of the flak batteries sited in the two towers of the dam scrambled to their battle stations and opened fire. Not a trace of breeze ruffled the surface of the black water, which was so still that it reflected the streams of tracer from the anti-aircraft guns like a mirror, leading Gibson and the pilots making the first bomb-runs to believe there were not three but six anti-aircraft guns firing at them.

At 0.28 on the morning of 17 May 1943 Guy Gibson's bomb-aimer released his bomb. Gibson banked round in time to see the yellow flash beneath the surface of the water and subsequent waterspout, but as the water splashed down and the smoke dispersed, it was clear that his bomb had failed to breach the dam. He then circled and called in the next pilot, John 'Hoppy' Hopgood, nonchalantly describing the task as 'a piece of cake'. It was anything but. Hopgood's Lancaster had already been hit by flak as they flew across Germany towards the target, wounding several of his crew and giving him a serious facial wound, but he flew on, pressing a handkerchief against the wound to staunch the flow of blood. Turning to evade the probing searchlight beams, he flew so low that he even passed beneath the high-tension cables of a power line stretching across a landscape as dark as the night sky above them.

As he made his bomb-run, Hopgood's aircraft was again hit by flak and burst into flames. Although his bomb was released, it was dropped too close and bounced clean over the dam wall. Although it destroyed the electricity power station at its foot, sending a column of thick, black smoke high into the air, it did no damage to the dam itself. Hopgood managed to coax his stricken and burning aircraft up to 500 feet, giving his crew a slim chance of survival, but as he shouted 'For Christ's sake, get out!'[5] the flames reached the main wing fuel tank, which exploded. Two crewmen did manage to bale out, and survived, to see out the war as PoWs. His badly wounded wireless operator also baled out, but his parachute did not open in time. Hopgood and the other three members of his crew were trapped in the aircraft as it crashed in flames, and all were killed instantly.

The twenty-five-year-old Australian pilot Harold 'Mick' Martin was next to make a bomb-run, with Gibson now bravely flying almost alongside him to distract the flak gunners and draw their fire. Martin's bomb was released successfully, but two flak shells had struck a petrol tank in his starboard wing and, though it was empty of fuel and there was no explosion, the impact may have thrown their bomb off course. As Martin flew on between the stone towers of the dam, the bomb hit the water at an angle and bounced away to the left side of the dam, before detonating in the mud at the water's edge. However, the force of the mud and water thrown up by the blast dislodged one of the flak guns on the southern tower from its mounting and the gun-crew of the battery in the northern tower were now also forced to change their red-hot gun-barrels.

In the lull in the firing, Melvin 'Dinghy' Young – so named after surviving two ditchings at sea – made his bomb-run. His Upkeep was released perfectly, bouncing in smooth parabolas along the surface of the water. It struck the exact centre of the dam and sank deep below the surface before detonating against the dam wall. The explosion sent up another huge waterspout, but once more it appeared to have failed to breach the dam.

The op looked to be heading for a disastrous failure, and the bouncing bombs' designer, Barnes Wallis, waiting in the operations room back in England – the basement of a large house on the outskirts of Grantham – was almost beside himself as the succession of coded signals came through, announcing the failure of each bomb-run. He had tested his original concept by bouncing his daughter's marbles across a tub of water, and had endured obstruction and ridicule before his ideas were accepted, but three bombs had already failed, and now the fourth bomb had also been dropped, once more without apparent effect.

Gibson now called up David Maltby, just twenty-three and, like Mick Martin, already the holder of the DFC and DSO. He was the fifth to make his approach, flying in at 60 feet above the water, with Gibson and Martin both acting as decoys to draw off some of the anti-aircraft fire. They even switched on their navigation lights to distract the flak-gunners, while their own forward gunners poured fire into the enemy positions, but although the raiders did not know this, all the anti-aircraft guns had now been put out of action and the gunners could only fire rifles at the Lancaster roaring across the lake towards them.

As he skimmed the surface of the water, his attention focused on the massive grey wall spanning the gap between the two towers, Maltby suddenly realised that the crest of the parapet was beginning to change shape. A crack had appeared, growing wider and deeper by the second as a section of masonry began to crumble and pieces of debris tumbled into the water. Close to his own bomb-release point, Maltby realised that Dinghy Young's bomb had made a small breach in the dam which was already beginning to fail.

As a result, Maltby veered slightly to port to target a different section of the dam just as his bomb-aimer dropped the Upkeep bomb. It was released perfectly, bouncing four times, before striking the dam wall and then sinking below the surface right against the dam face. Moments after Maltby's aircraft passed over the parapet, the bomb erupted in a huge column of water mixed with silt and fragments of rock.

It was still not clear at first if even this bomb had actually breached the dam, and, moments later, Barnes Wallis and the others listening in the operations room in England heard the terse radio signal: 'Goner 78A'. 'Goner' meant a successful attack, '7' signified an explosion in contact with the dam, '8' no apparent breach, and 'A' showed the target was the Möhne dam.

However, as the debris from the waterspout spattered down, Maltby could see that the dam – its masonry fatally weakened by the repeated bomb-blasts – was now crumbling under the monstrous weight of water it held, and as the raiders watched, the breach gaped wider. The torrent pouring through the ever-growing gap, dragging the anti-torpedo nets with it, hastened the Möhne dam's complete destruction. As it collapsed, widening into a breach almost 250 feet across, a wall of water began roaring down the valley, a tide of destruction sweeping away villages and towns in its path. Tragically, among the countless buildings destroyed were the wooden barrack blocks housing hundreds of East European women that the Nazis had compelled to work as forced labourers. Almost all of them were drowned, and they formed by far the largest part of the 1,249 people killed by the raid, one of the highest death tolls from a Bomber Command operation at that time.

Maltby sent a one-word radio transmission: the name of Gibson's black Labrador dog, which told all those waiting in the operations room that the Möhne dam was no more. Air Marshal Arthur Harris, the head of Bomber Command, turned to Barnes Wallis, shook his hand and said, 'Wallis, I didn't believe a word you said when you came to see me. But now you could sell me a pink elephant.'[6]

Gibson told Maltby and Martin, who had both used their bombs and sustained flak damage, to turn for home, while he, old Etonian Henry Maudslay, the baby-faced Australian David Shannon – only twenty, but another pilot who was already the holder of the DSO and DFC – and yet another Australian, twenty-two-year-old Les Knight, who 'never smoked, drank or chased girls',[7] making him practically unique in 617 Squadron on all three counts, flew on to attack the next target, the Eder dam. Thirteen storeys high, it was virtually unprotected by flak batteries, since the Germans believed that its position in a narrow, precipitous and twisting valley made it invulnerable to attack.

The approach was terrifying, a gut-wrenching plunge down the steep rocky walls of the valley to reach the surface of the lake, leaving just seven seconds to level out and adjust height, track and speed before releasing the bomb. Shannon and Maudslay made repeated aborted approaches to the dam before, at 01.39 that morning, Shannon's bomb-aimer finally released the Upkeep bomb. Shannon's angle of approach sent the bomb well to the right of the centre of the dam wall, but it detonated successfully and he was convinced that a small breach had been made.

Maudslay was next. Guy Gibson later described him pulling up sharply and then resuming his bomb-run, and other witnesses spoke of something projecting from the bottom of his aircraft, suggesting either flak damage or debris from a collision with the tops of the trees on the lake shore. Whatever the cause, Maudslay's bomb was released so late that it struck the dam wall without touching the water and detonated immediately, just as Maudslay was overflying the parapet. The blast wave battered the aircraft, and although Maudslay managed to give the coded message that his bomb had been dropped, nothing more was heard from him or his crew. He attempted to nurse his crippled aircraft back to England but was shot down by flak batteries on the banks of the Rhine near the German-Dutch border. The Lancaster crashed in flames in a meadow, killing Maudslay and all the other members of his crew.

Knight now began his bomb-run. Like Shannon, he made his approach over the shoulder of a hill, then made a sharp turn to port, diving down to 60 feet above the water, counted down by his navigator who was watching the twin discs of light thrown onto the black water by spotlights fitted beneath the fuselage, waiting until they converged into a figure '8' that showed they were at exactly the right height. Knight's bomb – the third to be dropped, and the last one the first wave possessed – released perfectly, hit the dam wall and sank before detonating. The blast drilled a hole straight through the dam wall, marked at once by a ferocious jet of water bursting from the downstream face of the dam. A moment later the masonry above it crumbled and collapsed, causing a deep V-shaped breach that released a tsunami-like wall of 200 million tons of water.

The remaining aircraft of the second and third waves were now making for the third and last target of the night, the Sorpe dam. It was of different construction from the other two – a massive, sloping earth and clay mound with a thin concrete core, rather than a sheer masonry wall – making it a much less suitable target for Wallis's bouncing bombs. Pilot Officer Joe McCarthy, a twenty-three-year-old New Yorker, was the only pilot of the second wave of five aircraft to reach the dam. Pilot Officer Geoff Rice had hit the sea near Vlieland, tearing his Upkeep bomb from its mounting and forcing him to abort the op and return to base. Like Astell, Flight Lieutenant Bob Barlow had collided with some electricity pylons, killing himself and all his crew, and Pilot Officer Vernon Byers had been hit by flak over the Dutch island of Texel. His aircraft crashed in flames, killing all seven men aboard.

New Zealander Les Munro was also hit by flak as he crossed the Dutch coast. As he reached his turning point approaching the island of Vlieland, he could see the breakers ahead and the sand dunes rising above the level of the sea, and gained a little height to clear them. He was, he recalled, flying 'pretty low, about thirty or forty feet, and I had actually cleared the top of the dunes and was losing height on the other side when a line of tracer appeared on the port side and we were hit by one shell amidships. It cut all the communication and electrical systems and everything went dead. The flak was only momentary, we were past it in seconds, but one lucky or unlucky hit – lucky it didn't kill anyone, or do any fatal damage to us or the aircraft, but unlucky that one shot ensured we couldn't complete the op we'd trained so long for' – had forced them to abort the op and turn for home.

For those who did make it to the target, the steep terrain and dangerous obstacles close to the Sorpe dam – tall trees on one side and a church steeple on the other – made it difficult to get low enough over the water for a successful bomb-drop. 'All that bombing training we'd done,' McCarthy's bomb-aimer, Johnny Johnson, recalls, 'we couldn't use because, as the Sorpe had no towers, we had nothing to sight on. Also, it was so placed within the hills that you couldn't make a head-on attack anyway. So we had to fly down one side of the hills, level out with the port outer engine over the dam itself so that we were just on the water side of the dam, and estimate as nearly as we could to the centre of the dam to drop the bomb. We weren't spinning the bomb at all, it was an inert drop.'

McCarthy made no fewer than nine unsuccessful runs before Johnson, to the clear relief of the rest of the crew, was finally able to release their bomb. Had the dam been as well defended by flak batteries as the other dams, it would have been suicidal to make so many passes over the target, but at the Sorpe, the main threat was the precipitous, near-impossible terrain surrounding it. It was an accurate drop and the bomb detonated right at the centre of the dam, blasting a waterspout so high that water hit the rear gun turret of the Lancaster as it banked away, provoking a shocked cry of 'God Almighty!' from the rear gunner.[8] McCarthy circled back but, although there was some damage to the top of the dam, there was no tell-tale rush of water that would have signalled a breach.

Two of the five aircraft from the third wave had already been shot down on their way to the target. Pilot Officer Lewis Burpee, whose pregnant wife was waiting for him at home, was hit by flak after straying too close to a heavily defended German night-fighter base and then hit the trees. His bomb detonated as the Lancaster hit the ground with a flash that his horrified comrades described as 'a rising sun that lit up the landscape like day'.[9] All seven members of the crew were killed instantly. Bill Ottley's Lancaster also crashed in flames after being hit by flak near Hamm. Ottley and five of his crew were killed, but by a miracle, his rear gunner, Fred Tees, although severely burned and wounded by shrapnel, survived and became a prisoner of war.

Ground mist spreading along the river valleys as dawn approached was now making the task of identifying and then bombing their target even more difficult, and the third aircraft of the third wave, flown by Flight Sergeant Cyril Anderson, was eventually forced to abort the op and return to base without even sighting the target.

Flight Sergeant Ken Brown, leading an all-NCO crew, did find the Sorpe, and after dropping flares to illuminate the dam they succeeded in dropping their bomb, once more after nine unsuccessful runs. Like McCarthy's, the Upkeep struck the dam wall accurately and also appeared to cause some crumbling of its crest, but the dam held.

Although the Sorpe still remained intact, for some reason the last aircraft, piloted by Bill Townsend, was directed to attack yet another dam, the Ennepe. Although he dropped his bomb, it bounced twice and then sank and detonated well short of the dam wall. Townsend's was the last Lancaster to attack the dams, and consequently so late in turning for home that dawn was already beginning to break, making him a very visible target for the flak batteries. A superb pilot – on his way to the target he had dodged flak by flying along a forest firebreak below the level of the tree-tops – he was at such a low level as he crossed the Dutch coast and flew out over the North Sea that coastal gun batteries targeting him had the barrels of their guns depressed so far that shells were bouncing off the surface of the sea, and some actually bounced over the top of the aircraft.[10]

Although the Sorpe dam remained intact, the destruction of the Möhne and Eder dams had already ensured that Operation Chastise was a tremendous success, but it had been achieved at a terrible cost. Dinghy Young's crew became the last of the night's victims. An unusual character with an interest in yoga, who used to 'spend much of his time during beer-drinking sessions, sitting cross-legged on tables with a tankard in his hand',[11] Young had reached the Dutch coast on his way back to base when he was shot down with the loss of all seven crew. His was the eighth aircraft to be lost that night, with a total of fifty-six crewmen killed or missing in action. Those waiting back in Lincolnshire for news, including Gwyn Johnson in her fitful sleep at her billet, faced a further anxious wait before the surviving Lancasters made it back. Townsend, the last to return, eventually touched down back at 617's base at Scampton at a quarter past six that morning, almost nine hours after the first of the Dambusters had taken off.

As usual, Gwyn Johnson had heard the aircraft taking off before she went to sleep the previous night and had woken up again as they were coming back. For reasons of security – and for Gwyn's peace of mind – Johnny hadn't told her about the op before taking off, and he didn't tell her he'd been part of the Dams raid at all until months after the event. 'I didn't really want to tell her I'd been on that particular op,' he says, 'as I suspected she might be annoyed I'd never mentioned it previously. Sure enough, when she did find out, she gave me an earful for not telling her in the first place!'

On the night of the raid, there had been 'no sleep for anyone' waiting back at Scampton as the hours ticked by. 'Our hearts and minds were in those planes,' said one of the WAAFs who were waiting to serve them a hot meal on their return. As the night wore on, twice they heard aircraft returning and rushed outside to greet crews who had been forced to turn back before reaching the target and were nursing their damaged aircraft home.

When they again heard engines in the far distance, the WAAFs were ordered back to the Sergeants' Mess to start serving food to the first arrivals. They waited and waited, but no aircrew came in. Two hours later, their WAAF sergeant called them together to tell them the heartbreaking news that out of nineteen aircraft that had taken off that night, only eleven had returned, with the presumed loss of fifty-six lives. (In fact fifty-three men had died;

the other three had been taken prisoner after baling out of their doomed aircraft.)

'We all burst into tears. We looked around the Aircrews' Mess. The tables we had so hopefully laid out for the safe return of our comrades looked empty and pathetic.' Over the next few days, the squadron routine slowly reasserted itself and the pain of those losses began to diminish, but 'things would never, ever be the same again'.[12]

The ground crews shared their sense of loss. 'The ground crews didn't get the recognition they deserved,' one of 617's aircrew says. 'Without them we were nothing. They were out in rain, snow and sun, making sure the aircraft was always ready, always waiting for us to come back. And when one didn't come back, it was their loss as much as anyone's.'[13]

The aircrews of 617 felt the deaths of their comrades and friends as keenly as anyone, of course. 'We had lost a lot of colleagues that night and there was a real sense of loss,' Johnny Johnson says. 'There were so many who didn't make it home – just a mixture of skill and sheer luck that it didn't happen to us as well.' However, most of the aircrews were veterans of many previous ops and, if not on this scale, had experienced the loss of friends a number of times before and developed ways of coping. It wasn't callousness, far from it, but with deaths occurring on almost every op they flew, men who dwelt on the deaths of comrades would not survive long themselves.

Losses of crewmates and friends were never discussed. 'It just never came up,' Johnny Johnson says:

though I did think about death when my roommate on 97 Squadron, Bernie May, was killed. We were on the same op together when his pilot overshot the runway on landing, went through a hedge and smashed the nose up. Bernie was still in the bomb-aimer's position and was killed outright. By the time I got back, all his gear had been cleared away from our room. It affected me that one minute he was there, and the next minute, no trace of him. Just bad luck really, but you just had to go on and find another friend. That was how it was then.

'The hardest part was writing to the relatives of those that didn't make it,' front gunner Fred Sutherland says. 'Trying to write to a mother, and all you could say was how sorry you were and what a good friend their son had been to you.'

'I'd lost friends and colleagues,' Johnson adds:

> *but never thought it would happen to me, and I had total trust in Joe McCarthy. He was a big man – six feet six – with a big personality, but also big in ability. He was strong on the ground and in the air, which gave the rest of the crew a tremendous boost. Joe had a toy panda doll called Chuck-Chuck, and we had a picture of it painted on the front of all the aircraft we flew. Other than that, I didn't believe in lucky charms – you made your own luck – but we had such confidence in Joe that it welded us together. We all gave him the best we could and trusted him with our lives and I never, ever, thought he'd not bring me back home.*

McCarthy was a genial giant who had spent some of his youth working as a lifeguard at Coney Island. After three failed attempts to join the US Army Air Corps, he crossed the border and joined the Royal Canadian Air Force instead. He came to England in January 1942 and flew on operations to the Ruhr even before he'd completed his advanced training. He then joined 97 Squadron in September 1942, where Johnny Johnston became his bomb-aimer. Most of the ops they flew together were to the ironically named 'Happy Valley' – the Ruhr, which had such formidable air defences that bombing there was anything but a happy experience, and many of their fellow aircrews lost their lives.

McCarthy led a multinational crew. He was from the Bronx in New York, his navigator Don McLean, rear gunner Dave Rodger, and flight engineer Bill Radcliffe were all Canadians, and the three Englishmen – Johnny Johnson, Ron Batson, the mid-upper gunner, and Len Eaton, the wireless operator – were NCOs.

The mixture of rank and nationalities, Johnson says, 'had no significance whatsoever to any of us. We were all on Christian-name terms, including Joe, and we all got on well. There was no stand-offishness, nothing to suggest any difference between any of us.' By contrast their first meeting with their new commanding officer had been chilly, but Gibson was already known to get on much better with men of his own class and background than with 'other ranks and colonials'. When Gibson was on 106 Squadron, Johnson says:

> he was known as the 'Arch-Bastard' because of his strict discipline, and one thing he didn't have was much of an ability to mix with the lower ranks; he wasn't able to bring himself to talk with the NCOs, and certainly not with the ground crews. He was a little man and he was arrogant, bombastic, and a strict disciplinarian, but he was one of the most experienced bomber pilots in Bomber Command, so he had something to be bombastic about. He spoke to us all at briefings, but he never spoke to me on a one-to-one basis, or ever shook my hand, or even acknowledged me. But that's just the way he was and he was a true leader in the operational sense, his courage at the dams showed that.[14]

Gibson expected others to show no less courage and dedication, and he could be abrupt, even merciless, with those whom he decided had failed to meet his exacting standards. One of the reserve crews on the Dams raid, piloted by Yorkshireman Cyril Anderson, had been redirected to the Sorpe dam from their original target, but failed to find it. After searching for forty minutes, suffering a mechanical problem, and with dawn already beginning to lighten the eastern sky, Anderson aborted the op and returned to base with his Upkeep bomb still on board. Whether or not Anderson's humble Yorkshire origins played a part in Gibson's decision, he showed his displeasure by immediately posting Anderson and his crew back to 49 Squadron. In any event, the commander of that squadron did not share Gibson's opinion of Anderson, and in fact recommended him for a commission as an officer shortly after he rejoined the squadron.[15]

However, others, even among the 'other ranks', found Gibson easier to deal with. He and his beer-drinking black Labrador – sadly run over outside Scampton the night before the Dams raid – were regulars in the Officers' Mess, and wireless operator Larry Curtis, whose Black Country origins and rise from the ranks would not have made him a natural soulmate for Gibson, said of him: 'I know some people said he was a bit hard, but I got on well with him … I found him hard but very just, you couldn't ask any more than that. When it was time for business he was very businesslike, when it was time to relax, he relaxed with the best of them. I only regret I never had a chance to fly with him, because he was a wonderful pilot.'[16] Gibson had demonstrated his skill and bravery as both a pilot and a leader many times, but the Dams raid was to be his crowning achievement.

Bomber Command C-in-C Arthur Harris had argued forcefully against the raid beforehand, describing the idea as 'tripe of the wildest description',[17] but he and Air Vice Marshal The Honourable Ralph Cochrane, commander of 5 Group, of which 617 Squadron formed a part, had hurried from Grantham to Scampton to congratulate the returning heroes. For them, as for the government, the press and the nation, starved for so long of good news about the war, any reservations about the aircrew losses were swept away in the jubilation about the Dams raid's success. 617 Squadron had shown Hitler and his Nazi hierarchy that the RAF could get through and destroy targets they had previously thought invulnerable. They had dealt a severe blow – albeit a short-term one, since the dams were repaired within months – to German arms production. They had forced the Germans to divert skilled workmen from constructing the Atlantic Wall to repair the dams, and that might well have a significant impact on the chances of success of an Allied invasion of France when it eventually came.

Even those successes paled beside the huge impact the raid had on the morale of the people of Britain, and on public opinion around the world, particularly that of our sometimes reluctant and grudging ally, the United States. 'I don't think we appreciated how important the raid was in that respect,' Johnny Johnson says, 'until we saw the papers the next morning, when it was plastered all over the headlines. There had been the victory at El Alamein a few months before and now this, and it was a big, big change in what had been a bloody awful war for us until then.'[18]

Ironically, Cochrane, a severe-looking man with a high forehead and a piercing stare, had warned the crews at their final briefing that the Dams raid might be 'a secret until after the war. So don't think that you are going to get your pictures in the papers.'[19] Security before the raid had been so tight that one of the local barmaids in Lincoln was sent on holiday, not because she was suspected of treachery but because she had such a remarkable memory and such a keen interest in the aircrews that it was feared she might inadvertently say something that would compromise the op.

However, once the raid was over, any considerations of the need for secrecy were swept away like the dams, by the propaganda value of publicising the raid. The Dams raid chimed perfectly with the narrative created by British propagandists: the plucky but overwhelmingly outnumbered underdog fighting alone, and through expertise, ingenuity, courage and daring, breaching the defences of the monolithic enemy.

The *Daily Telegraph* exulted on its front page that 'With one single blow, the RAF has precipitated what may prove to be the greatest industrial disaster yet inflicted on Germany in this war,'[20] and the other newspapers were equally triumphant in tone. Guy Gibson was awarded a Victoria Cross for his leadership of the raid and more than half the surviving members of the squadron were also decorated, but Air Marshal Arthur 'Bomber' Harris's euphoria in the immediate aftermath of the raid soon gave way to pessimism. In a letter to the Assistant Chief of Air Staff he said he had 'seen nothing … to show that the effort was worthwhile, except as a spectacular operation',[21] and although he often appeared unmoved by aircrew losses on Main Force – the major part of Bomber Command which carried out the near-nightly area bombing of German cities – he later remarked that missions where Victoria Crosses went along with high losses should not be repeated.

However, the damage to German industry and infrastructure was undoubtedly considerable. Thousands of acres of farmland

and crops were buried by silt, and the land could not be tilled for several years. Food production, output from the coal mines and steel and arms manufacture were all badly hit. The destruction of power stations further reduced industrial output, and the disruption to water supplies caused by wrecked pumping stations and treatment plants not only deprived factories of water but left firemen in the industrial towns unable to extinguish the incendiaries dropped by RAF bombers. The destruction of 2,000 buildings in Dortmund was directly attributed to this.[22]

Nonetheless, when Barnes Wallis, who had also hurried to Scampton to add his congratulations, was told of the deaths of so many airmen, he cried. 'I have killed all those young men,' he said. His wife wrote to a friend the next day: 'Poor B didn't get home till 5 to 12 last night ... and was awake till 2.30 this morning telling me all about it ... He woke at six feeling absolutely awful; he'd killed so many people.'[23] Although Guy Gibson and the others did their best to reassure him, according to his daughter Mary, Wallis lived with the thought of those deaths throughout the rest of his life.[24]

The Australian pilot Mick Martin's opinion was that most of the casualties on the Dams raid had been caused because their navigation had been 'a bit astray' and, more importantly, because pilots didn't keep below 200 feet on their way to the target. 'Of course,' said his rear gunner, who shared Martin's views, 'it's easy to criticise when you're sitting down at the back, arse about face, instead of up front where you see things rushing at you head-on, and even though it is moonlight, which can be almost as bright as day, there is a great risk of miscalculation of height, and objects such as power lines don't show up like the Boston Stump'[25] – the remarkably tall tower of St Botolph's church in Boston, rising over 270 feet above the Lincolnshire fens, and visible from miles away.

The surviving aircrews washed away the bad memories of the night and toasted their success with a few beers. The Mess bars were reopened and drink was taken in considerable quantities. Some of the aircrews had 'a mug of beer in each hand', according to the Squadron Adjutant, Harry 'Humph' Humphries. 'I could hardly keep my eyes open, yet these lads drank away as though they had just finished a good night's sleep. The only thing that contradicted this was the number of red-rimmed eyes that could be seen peering into full beer mugs.'[26] 'The celebrations were in full swing,' Les Munro says, 'and I wondered if I really deserved to be part of it all. Should I be taking part in this when I hadn't reached the target? All that training, preparation had been for nothing. But no one ever made any comment about us not getting to the target.'

The celebrations ended with a conga line of inebriated airmen visiting Station Commander Group Captain Charles Whitworth's house and then departing again, having deprived the sleeping Whitworth of his pyjamas as a trophy. But while the crews had been celebrating or drowning their sorrows, Harry Humphries had the painful task as Squadron Adjutant of composing fifty-six telegrams and then fifty-six personal letters to wives and families, telling them that their loved ones would not be coming home.

Once the initial euphoria had passed, reactions within the squadron to the success of the Dams raid tended to be more muted. Some were elated but for most, 'we were satisfied in doing the Dams raid,' Johnson says, 'but nothing more really. It was just another job we had to do. I had no sense of triumphalism or excitement, we realised what we had done, and were very satisfied, but that was all.'[27] However, in a public demonstration of the importance of the raid to British morale, King George VI and Queen Elizabeth paid a visit to the squadron on 28 May 1943, with all the aircrews lined up, standing to attention and with their toecaps touching a white line that had been specially painted on the grass. They spent five hours there, though 'naturally, of course, neither the King nor Queen visited the Sergeants' Mess'.[28]

The gallantry awards for the raid – one Victoria Cross, five DSOs, fourteen DFCs, eleven DFMs and two Conspicuous Gallantry medals – were then awarded by the Queen in an investiture at Buckingham Palace on 22 June. Johnny Johnson hadn't 'the foggiest idea' what the Queen said to him. 'I was so bloody nervous. All I can remember saying is "Thank you, Ma'am" and that was it.'[29] Although he was a teetotaller – 'I couldn't even stand the smell of beer,' he says, 'so I never went into the bars apart from a quick dash at lunchtime to pick up cigarettes'[30] – he was in a minority of close to one on the squadron, and his fellows more than made up for him, with the investiture the trigger for a marathon booze-up. They were given a special carriage on the train taking them to London, where they had 'a high old time', and after chatting to the driver and his fireman during a stop at Grantham, Mick Martin's crew made the next stage of the journey down to Peterborough on the footplate of the engine. They each threw a few shovelsful of coal into the firebox and took turns to operate the regulator under the supervision of the driver and 'gave the old steam chugger full bore'. The driver offered to stand them a drink at any pub of their choice in London, and they duly obliged him at the appropriately named Coal Hole in the Strand.

The more well-to-do of the aircrews stayed at the Mayfair Hotel, while the less affluent settled for the Strand, and in the words of one of the Australians, 'for twenty-four hours there was a real whoopee beat-up'.[31] As a result none of them were looking at their best on the morning of the investiture. Mick Martin's braces had disappeared somewhere along the line, but his resourceful crewmates 'scrounged a couple of ties and trussed him up like a chicken, sufficient to get him onto the dais at Buckingham Palace, down the steps and off again'.

The squadron's resourcefulness was also demonstrated by one of Mick Martin's crewmen, Toby Foxlee, who, as petrol was strictly rationed, obtained a regular supply of fuel for his MG by 'a reduction process' which involved pouring high-octane aircraft fuel through respirator canisters he had scrounged; 'the little MG certainly used to purr along pretty sweetly'.[32]

On the evening of the investiture, A. V. Roe, the manufacturers of the Lancaster, threw a lavish dinner party at the Hungaria restaurant in Regent Street for the 'Damn [*sic*] Busters following their gallant effort on the Rhur [*sic*] Dams'[33] – the budget for the dinner clearly didn't extend to a proofreader for the menu! Given the austerity of wartime rationing, a menu including 'Crabe

Cocktail, Caneton Farci a l'Anglaise, Asperges Vertes, Sauce Hollandaise, and Fraises au Marasquin', washed down with cocktails and ample quantities of 1929 Riesling, 1930 Burgundy and vintage port, was an astonishing banquet.

As the heroes of 617 Squadron celebrated their awards that night, Bomber Command's war against Germany ploughed on with a raid by nearly 600 aircraft against the city of Mulheim, north of Düsseldorf. The city's own records describe the accuracy of the bombing and the ferocity of the fires. Roads into and out of the area were cut and the only means of escape was on foot. The rescue services were overwhelmed, resulting in terrible destruction. Five hundred and seventy-eight people were killed and another 1,174 injured. Public buildings, schools, hospitals and churches were all hit. The German civilian population was paying a high price for Nazi aggression.[34]

Churchill's War Cabinet were quick to use the success of the Dams raid in the propaganda war against the enemy, and some of 617's aircrews were sent on publicity tours that spanned the globe, spreading the message about how good the RAF's premier squadron was through the world's press, radio stations and cinema newsreels. The main attraction was of course the leader of the raid, Guy Gibson, who was taken off ops, stood down as commander of the squadron, and sent on a near-permanent flag-waving tour, though before he left he managed to fit in a last trip to see 'one of the local women, a nurse, with whom he had been involved while his wife was in London'.[35] Gibson also 'wrote' (it is possible that Roald Dahl, working as an air attaché at the British Embassy in Washington, was the actual author) a series of articles, and a draft script for a movie of the Dams raid that director Howard Hawks was contemplating, then settled down to write his account of the Dams raid and his RAF career: *Enemy Coast Ahead*.

Mick Martin was by now widely recognised as the finest pilot on the squadron and had a wealth of operational experience behind him, but either his relatively lowly rank of Flight Lieutenant, requiring a double promotion to get him to the rank of

Wing Commander – 'it was considered not the done thing for him to jump two ranks'[36] – or a belief among his superiors that he was a lax disciplinarian had counted against him – or perhaps the RAF 'brass' just didn't want an Australian as CO. In any case, after just six weeks, he was replaced by Squadron Leader George Holden, who had taken over Guy Gibson's crew following his departure. If Holden shared something of Gibson's arrogance and his coolness towards NCOs, he lacked the former commander's charisma and leadership ability, and was not generally popular with his men.

His reign as commander of 617 was not to prove a long one, and in the perhaps biased opinion of Mick Martin's rear gunner, 'somebody made a bad mistake' in appointing Holden at all, since his knowledge of low flying was 'practically nil'.[37] Martin, by contrast, was 'superb, a complete master of the low-flying technique', and so dedicated that he continued to practise his skills every day. One new recruit to the squadron was given Martin's formula for success at low level: 'Don't, if you can help it, fly over trees or haystacks, fly alongside them!'[38] His bomb-aimer, Bob Hay, was also the Squadron Bombing Leader, charged with ensuring his peers achieved the highest possible degree of accuracy.

In this, he was encouraged by Air Vice Marshal Ralph Cochrane, who on one celebrated occasion even flew as bomb-aimer with Mick Martin's crew on a practice bombing detail. Cochrane arrived 'all spick and span in a white flying suit', took the bomb-aimer's post and achieved remarkable accuracy. The results were shown to the other crews, with the implicit message that if an unpractised bomb-aimer like Cochrane could achieve such results, the full-time men on 617 Squadron should be doing a lot better themselves.

While Martin was in temporary charge, his crew had put their enforced spare time to good use by creating a garden in front of 'the Flights' – the place where they spent their time before training flights or ops, often hanging around waiting for the weather to clear. Having found a pile of elm branches, they erected a rustic fence with an arch at the front and the squadron number at its apex, picked out in odd-shaped pieces of wood. They scavenged, dug up and, in cases of dire necessity, bought plants and shrubs, creating a peaceful haven. Sitting there in the sunshine, inhaling the scents of flowers and listening to the birdsong and the drowsy sound of bees, they could imagine themselves far from the war ... until the spell was broken as the Merlin engines of one of the squadron's Lancasters roared into life.

As the euphoria of the Dams raid faded, the remaining men of 617 and the new arrivals brought in to replace the crews lost on the raid settled down to a period of training which extended, almost unbroken, for four months. 'There was quite a gap between ops after the dams,' Fred Sutherland says, 'though I must say, I wasn't that bothered, and in no hurry to get back on ops.' Not all of the replacements were as experienced as the original crews. When Larry Curtis first reported to his commander, Gibson looked him over and then said, 'I see you haven't got any decorations, which surprises me.'

'The day after I joined, I was on operations,' Curtis said, with a grin, 'which surprised me!'[39]

Born at Wednesbury in 1921, Curtis joined the RAF Volunteer Reserve from Technical College in November 1939. After training as a wireless operator/air gunner, he was posted to 149 Squadron, flying Wellingtons, and took part in the attack on the German battle cruisers *Scharnhorst* and *Gneisenau* at Brest. After completing thirty ops, Curtis was sent to a bomber training unit as a wireless instructor, but still flew on the first thousand-bomber raid to Cologne. Commissioned in January 1943, he converted to the Halifax and joined 158 Squadron. During a raid on Berlin in March 1943 his aircraft was hit by flak over the target and all four engines stopped. As the aircraft spiralled towards the ground, the crew were about to bale out, when one engine picked up and the pilot was then able to recover and start the remaining three. Having already completed two bomber tours, Curtis joined Mick Martin's crew on 617 Squadron in July 1943 as one of the replacements for the men lost on the Dams raid.[40]

One of the main attractions for a lot of the aircrew joining 617 Squadron was the chance to fly low level. After flying at 10,000 to 15,000 feet on his previous ops, the thought of operating at 100 feet or even less was thrilling for Johnny Johnson. 'I thought, Wahey! Before I had been sitting on top of those bloody clouds and you could see nothing until you got to the target and then all you saw was rubbish. So it was absolutely exhilarating, just lying there and watching the ground going Woooof! Woooof! Wooooof! underneath me!' Against the regulations, he always took off and landed from his position in the nose. 'I didn't see the point of trying to stagger back and forwards to the normal position by the spar. So I could see the runway rushing beneath my nose every time we took off, and as we landed, the runway raced up to meet me – I loved that!'[41] One of their practice routes was over the Spalding tulip fields, and Johnson remembers a guilty feeling as they flew over the fields at low level, leaving a snow-storm of multi-coloured, shredded blooms and petals in their wake, torn up by the slipstream.[42]

In some ways the tail gunners had the worst job, with the greatest amount of time to ponder their fate. One rear gunner found it:

a very lonely job, cold and lonely, stuck out at the back of the aircraft with no one to glance at for reassurance or a little comfort. The first op was the worst; the only person who knew what it would be like was the skipper, because he had been on one op before with another crew. And he wouldn't tell us what it was going to be like; he didn't want to put the wind up us. We were naive and quite happy to trust to luck. Oh, I was afraid, especially of what was to come, the unknown, but you just couldn't show it.

The odds were stacked against us, we knew that at the time, knew that the losses were huge. It's not a particularly nice thought, but once you were in that situation you just had to carry on and do the job you were trained for, and you just blotted out everything else. You ignored the flak, ignored the aircraft going down around you. What else could you do? In many ways you were stuck in the middle of that horror, those losses. You don't 'cope' with it, do you? You just do it.[43]

Having been specifically formed for that single op against the dams, it now appeared to pilots like Les 'Happy' Munro (also known as 'Smiler' – both nicknames were sarcastic, because he was famed for his dour demeanour) that 'nobody knew what to do with us. There was a hiatus, a sense of frustration. What was 617 Squadron for? The powers that be couldn't seem to make up their minds about what to do with this special squadron they'd created.' Munro was a New Zealander, but had enlisted because of:

a general feeling that we were part of the British Empire, and had an alle-giance to King and Country. We were really aware, through radio broadcasts and cinema newsreels, of what Britain was facing and what they were being subjected to: the air attacks, the Blitz. It was a sense of duty for most of us young men in New Zealand to fight for 'the old country' against the Nazis, but of course I had no idea at all of what would happen to me or what was to come: the devastation, the dangers, the losses I'd see and experience. Nor that seventy years on I'd still be talking about it!

'Being on 617 meant that there could be quite a long gap between ops,' another crewman says.

I remember going on leave and meeting friends from my engineers' course who had nearly finished their thirty-op tour whilst I'd only done three or four. That's when you began to understand how different 617 was. My friends said I was a 'lucky bastard' for being on 617 and not Main Force, and they were probably right – I wouldn't have liked to be on MF from all the stories I heard and read afterwards.[44]

617's lack of ops led aircrew from 57 Squadron, also based at Scampton, to shower them with jibes and insults and give them the sardonic nickname 'The One-op Squadron'. The men of 617 retaliated by 'debagging many of the Fifty-Seven men in a scragging session in the Officers' Mess',[45] but they also ruefully acknowledged their reputation in their own squadron song, which they sang to the tune of a hymn written in 1899, 'Come and Join Our Happy Throng':

The Möhne and Eder dams were standing in the Ruhr,
617 Squadron bombed them to the floor.
Since that operation the squadron's been a flop,
And we've got the reputation of the squadron with one op.

Come and join us,
Come and join us,
Come and join our happy throng.

Selected for the squadron with the finest crews,
But the only thing they're good for is drinking all the booze.
They're not afraid of Jerry and they don't care for the Wops,
Cos they only go to Boston to do their bloody ops.

Come and join us …

To all you budding aircrew who want to go to heaven,
Come join the forces of good old 617.
The Main Force go to Berlin and are fighting their way back,
But we only go to Wainfleet where there isn't any flak.

Come and join us …
Come and join our happy throng.[46]

CHAPTER 2

What Next?

While some crewmen on 617 Squadron were chafing at their inactivity, Johnny Johnson welcomed the lack of ops. 'It meant I had more time with Gwyn, and we had so little time together that it was important to make the most of every minute.'[1]

Born in 1921 near Horncastle in Lincolnshire, Johnson had been one of six children.

Unfortunately, my mother died two weeks before my third birthday, so I never really knew a mother's love. It really affected me – I remember seeing her in the hospital bed. I was standing next to my father and another man, and my father described me to him as 'this one is the mistake'. I remember that to this day.

I had a very unhappy childhood. He wouldn't let me go to grammar school and was ruining my life. Eventually I went to the Lord Wandsworth agricultural college for children who had lost a parent. Again, my father had said no, but the local squire's wife went to see him and told him in no uncertain terms that he had to let me go! I was eleven at the time.

By November 1940 Johnson was a trainee park keeper with ambitions to be the superintendent of a big London park, but with London suffering under the Blitz, he thought: 'What the hell am I doing here?' He wanted to be part of the war, not left behind, but didn't want to join the Army. 'I had seen the reports of World War One trench warfare, casualties and the like, and didn't want any of that, and I didn't like water, so the Navy was out! So that left the RAF. I wanted to be on bombers so I could take the war to the enemy, to get at the Germans. I had no thought of any dangers back then, I just didn't think about it.'

Like many other British aircrew, Johnson did his initial training in America because, even before Pearl Harbor in December 1941 brought the Americans into the war, the US government had arranged discreet support for the British war effort by secretly training British aircrew under the Arnold Scheme. To maintain the fiction of American neutrality, aircrew wore civilian clothes and travelled via Canada, before slipping across the US border.

Johnny Johnson pictured in 1947

Johnson returned to the UK in January 1942 and, desperate to get into action, volunteered to train as a gunner – the shortest training course. Testing his resolve, the president of the selection board said to him, 'I think you'd be afraid to be a gunner, Johnson.'

'I don't think so, sir,' he said. 'If I was, I wouldn't have volunteered in the first place!' 'So I gave as good as I got,' he says now, with a chuckle. 'I was going to prove to him that I wasn't afraid! I had no sense of fear or thoughts of what the future might hold, and certainly no idea of the losses Bomber Command would suffer.'

Johnson retrained as a bomb-aimer, not least because they earned five shillings (25p – about £10 at today's values) a day more than gunners. As a bomb-aimer, he manned the front gun turret on the route out and only went into the bomb-aimer's compartment as they approached the target. He then fused and selected the bombs, set the distributor and switched on his bombsight. Lying in the nose of the aircraft on the bombing run, he could see the flak coming up at him, but had to ignore that and concentrate on doing his job.

From a distance the flak bursts could seem almost beautiful, opening like white, yellow and orange flowers, but closer to, dense black smoke erupted around them and there was the machine-gun rattle of shrapnel against the fuselage and the stench of cordite from each smoking fragment that pierced the aircraft's metal skin.

'I don't think I was afraid,' he says:

but when you see the flak you have to go through, I think anyone who didn't feel some apprehension was lacking in emotion or a stranger to the truth, but you didn't want to let anyone down. The crew were doing their jobs and mine was to get those bombs on the target to the exclusion of all else. Once we got to the target area, I was too busy concentrating on the bombsight and dropping the bombs in the right place to worry about what else was going on.

Despite his initial scepticism about the value of a 'special squadron', in mid-July 1943, two months after the Dams raid, Bomber Harris proposed using 617 Squadron to assassinate the Italian dictator, Benito Mussolini. A letter to the Prime Minister from the Chief of the Air Staff revealed that Harris had asked permission to bomb Mussolini in his office in the Palazzo Venezia in Rome, and his house, the Villa Torlonia, simultaneously, 'in case Il Duce is late that morning … Harris would use the squadron of Lancasters (No. 617) which made the attacks on the dams. It is manned by experts and is kept for special ventures of this kind.' It was suggested that if Mussolini were killed 'or even badly shaken', it might increase the Allies' chances of speedily forcing Italy out of the war. However, the plan was vetoed by Foreign Office officials, who were unconvinced that eliminating Mussolini would guarantee an Italian surrender and feared that it might even lead to his replacement by a more effective Italian leader.[2]

Two days later, on 15 July 1943, 617 Squadron at last saw some fresh action, though it proved to be what one Australian rear gunner dismissed as 'a stooge trip' – an attack on a power station at San Polo d'Enza in northern Italy. 'We screamed across France at practically zero level, climbed like a bat out of hell to get over the Alps, and then screamed down on to St Polo and completely obliterated the unfortunate power station without seeing a single aircraft or a single burst of flak.'[3] Other crews would have been grateful even for that level of activity, one pilot complaining that after two months' inaction, when they finally did get an op it was 'to bomb Italy ... with leaflets'. As Joe McCarthy grumpily remarked, it was 'like selling god-damned newspapers'.[4]

There was only one thing McCarthy hated more than dropping leaflets, and that was signing forms, and one of his duties was to sign his aircrews' logbooks every month. It was a task he seemed to find more difficult and intimidating than the most dangerous op. His education had been as much on the streets of the Bronx and the beaches of Coney Island as in the classroom, and his handwriting was laborious and painfully slow. He would put the task off as long as possible and when he could finally avoid it no longer, his crewmates would gather to watch, in fits of laughter at the sight of their huge and normally unflappable Flight Commander, with his tongue protruding from the corner of his mouth, sweating buckets and cursing under his breath as he struggled to complete the hated task.

During that summer of 1943, 617 Squadron moved from Scampton to Coningsby, where they would have the advantage of concrete runways, rather than the grass strips they had been using at Scampton. Those grass runways, camouflaged with 'hedgerows' painted on the turf to fool German raiders, had been less of a problem than they might have been, because the airfield was at the top of an escarpment and the natural drainage prevented Scampton from becoming boggy in all but the most relentless wet weather. However, with the squadron's Lancasters carrying increasingly heavy fuel- and bomb-loads, a move to Coningsby was necessary, and 617's pilots were soon airborne and familiarising themselves with the local landmarks there: a windmill in the nearby Coningsby village, Tattershall Castle to the north-west, beyond the river Bain, and, most distinctive of all, the towering St Botolph's church, universally known as the 'Boston Stump'. It was a rheumy, water-filled land, criss-crossed by dykes and ditches, and prone to autumn mists and winter fogs that often forced returning aircraft to divert elsewhere. There were farms dotted among the heathland and birch woods, rich pastures and water meadows, but to many of the aircrew the endless plains beneath the vast canopy of the skies seemed echoingly empty of life.

★   ★   ★

During the summer of 1943, Main Force had continued to take the war to the enemy, with Operation Gomorrah – the virtual destruction of Hamburg in a raid beginning on 24 July – creating havoc on an unprecedented scale. In one hour alone, 350,000 incendiaries were dropped there, and succeeding waves of British and US bombers over the next few days created firestorms that engulfed the city, killing 30,000 people. Elsewhere in the war, the tide was increasingly running in the Allies' favour. The Battle of Kursk had been launched by the Nazis in early July, but it proved to be their last major offensive on the Eastern Front, and the Soviets first neutralised the attack and then launched their own counter-offensive, driving the Germans back. In the west, the invasion of Sicily began on 10 July, and within five weeks the whole island was under Allied control, while on the Italian mainland Mussolini was deposed on 25 July 1943.

617 Squadron's long period of relative inactivity came to an abrupt end on 14 September 1943, when they were tasked with attacking the Dortmund–Ems Canal, a waterway 160 miles long, and the only one linking the Ruhr valley with eastern Germany and the ports of the Baltic and North Seas. That made it the most important canal system in Germany, a vital artery feeding Germany's war industries with strategic materials including the crucial

imports of Swedish iron ore, and transporting finished products that ranged from arms and munitions to prefabricated U-boat sections.

The canal was most vulnerable north of Münster around Ladbergen, where it ran in twin aqueducts over the river Glane. To either side of the aqueducts the canal was carried in embankments raised above the level of the surrounding land, and these, rather than the aqueducts, were designated as 617 Squadron's targets with the first operational use of much more powerful 12,000-pound High Capacity (HC) bombs, of which three-quarters of the weight was high explosive, compared with half in the smaller bombs.

On the face of it, 617's task was simple: bombing from 150 feet at a speed of 180 miles an hour, they were to drop their bombs on a precise aiming point within 40 feet of the west bank of the canal until a breach had been achieved. The remaining bombs were then to be dropped on alternate banks of the canal, moving north at 50-yard intervals to ensure as widespread a destruction of the canal embankments as possible. Even one bomb breaching the embankment would drain the canal, halting the flow of barge traffic, flooding the surrounding area and preventing millions of tons of Nazi supplies and weapons of war from reaching the front lines. However, the HC bombs were like elongated dustbins, built without streamlining and only small fins to enable them to fit into the bomb-bay. This made them unstable in flight and hard to drop accurately.

Six Mosquito fighter-bombers were to escort the squadron's Lancasters, operating as 'can-openers' by dealing with any flak hot-spots on the route. The Lancasters were to approach the target at extreme low level – 30 feet over Holland and Germany – before climbing to 150 feet to bomb. Although they had been practising for several weeks, flying low level along English canals, first by day and later by night, not everyone was happy with the idea of another low level attack on a heavily defended target. 'Our losses at the dams had been around fifty per cent,' Fred Sutherland says. 'And certainly I had doubts about this next op. My main concern was flying around at night, at very low level, with all those power cables criss-crossing everywhere.'

Born in 1923, Sutherland was Canadian, a full Cree Native American, who had volunteered for the Royal Canadian Air Force the minute he turned eighteen. 'I couldn't wait to get in the war in any way possible,' he says. 'Everybody wanted to get in. We were still suffering from the Depression, unemployment was high and it was a means to escape all that. All the talk was about the war and I wanted to be involved. I didn't really understand what it would be like though, I had no idea what was to come, what I'd go through, so I suppose I was naive.' After completing an air gunner's course, he crewed up with Les Knight, a 'short but very muscular' Australian pilot, 'strong in the shoulders and arms. He was a wonderful pilot,' Sutherland says, 'very quiet, but if you were out of line, he quietly told you that you'd better not do that again.'

David Maltby's crew

One of the two four-ships – formations of four aircraft – making the raid was to be led by Squadron Leader David Maltby, a very skilful pilot who had completed thirty ops over Germany and been awarded a Distinguished Flying Cross even before joining 617. Still only twenty-three, having struck the fatal blow against the Möhne dam, he was now one of the most highly decorated officers in the RAF and, like his comrades on the Dams raid, a national celebrity. A fun-loving, gentle giant over six feet tall, he was the life and soul of every party and always up for a prank; while training for the Dams raid, he had often 'buzzed' his wife Nina's family farm. His first child, a son, had been born soon afterwards; the shock of Maltby's aerobatics overhead may or may not have hastened the birth.

Maltby's personal good luck token was a filthy, oil-stained forage cap. He had worn it on the night of the Dams raid, never flew without it, and even wore it on parade. He donned it once more as he prepared to lead Operation Garlic – the raid on the Dortmund–Ems Canal. It would be his crew's first operation since the dams, and they would be flying at low level, straight into anti-aircraft gunfire, just as they had at the dams. The op was scheduled as a night-time raid on 14 September 1943, and eight Lancasters took off around midnight, but were recalled within forty minutes because of cloud obscuring the target. However, that 'boomerang' order resulted in tragedy. After acknowledging the order to return to base, Maltby's Lancaster crashed into the North Sea 8 miles off Cromer on the Norfolk coast, killing everyone on board. Famed as the man who breached the Möhne dam, Maltby had now joined the mounting tally of the squadron's dead.

Although the official accident report mentioned 'some obscure explosion and a fire' before the aircraft's fatal crash, it was believed for many years that Maltby had simply misjudged his height and dipped a wing into the sea, with fatal consequences. However, a rival theory has recently been advanced, claiming that he collided with a Mosquito from 139 Squadron that was returning from a separate raid on similar routing, and was also lost without trace that night.[5]

Dave Shannon circled over the crash site for over two hours until an air-sea rescue craft arrived, but Maltby's body was the only one ever to be recovered; his fellow crew members were lost in the depths of the North Sea, and are listed simply as having 'no known grave'. Their average age was just twenty. David Maltby is buried in a quiet corner of St Andrew's churchyard at Wickhambreaux near Canterbury, the church in which he had married Nina just sixteen months earlier. His son, just ten weeks old at the time of Maltby's death, would now never know his father.

The following night, the surviving members of the squadron returned to the Dortmund–Ems Canal, with Mick Martin's crew replacing Maltby's. 'Crews as a whole accepted the loss of a friend as a downside of the war in the air they were engaged in,' Les Munro says, a view echoed by Larry Curtis. 'One accepted the fact that you weren't coming back from this war. I did and most people did and it helped a lot. You were frightened but you knew it had to be done, so you did it.'[6] However, Maltby's death had given some of the aircrews pause for thought, and there was considerable trepidation about the op. 'I knew Dave Maltby,' Johnny Johnson recalls. 'He and Les Munro were on 97 Squadron with us, so when he'd been lost the night before, there was already a sense of it being a dodgy op.'[7]

The nervousness about the op only served to strengthen the importance of the pre-flight rituals or superstitions that almost all the crews followed. Every aircraft was carrying eight men rather than the usual seven, with an extra gunner aboard to ensure that all the gun turrets would be manned at all times throughout the flight. Unlike most of the other crews, one of Les Knight's crew's rituals was *not* peeing on the rear wheel before boarding their aircraft, and when Les Woollard, the extra gunner they were carrying, began to do so they all rushed over and pulled him away. 'We were not really bothered though,' Knight's front gunner, Fred Sutherland says, 'we were just fooling with him.' Sutherland thinks his crew was not superstitious, but then adds, 'but we always did the same routine. I always ate my chocolate bar when we were charging down the runway, and I always wore the same socks that my girlfriend, now my wife, had knitted for me.'

Eight Lancasters, all carrying 12,000-pound delayed-action bombs, and flying in two four-ships, took off at midnight on a beautiful, clear moonlit night. As they crossed the North Sea, the lessons learned on the Dams raid led them to adopt a new method of crossing the hostile coastline. Instead of flying a constant low level approach, they climbed before reaching the coast, but then went into a shallow dive back to low level, building up speed before flashing over the coastal flak batteries. 'And it really was

low level,' wireless operator Larry Curtis recalls. 'I can remember
the pilot pulling up to go over the high-tension cables.'[8] 'What
was really scary for me were the power wires,' adds Fred Sutherland, who had the closest view of them from his gun turret under
his aircraft's nose. 'Even if there was moonlight, you couldn't see
the wires until you were practically on them, and once you hit
them, that was it, you were done for.'

George Holden led the formation, with Mick Martin on his
starboard flank and Les Knight to port. Rear gunner Tom Simpson heard Martin and his bomb-aimer, Bob Hay, complaining
that Holden was flying too high, allowing the searchlights to pick
them up. 'We seemed to be getting into a lot of trouble and I had
never experienced such intense ground fire.' Just before reaching
the small German town of Nordhorn, Martin was, as usual, flying
'lower than low', squeezing between some factory chimney
stacks, 'the top of the stacks being higher than we were', but
Holden, still much higher, was drawing heavy anti-aircraft fire.
Fred Sutherland, front gunner in Les Knight's crew, whose job
was to attack 'any ground flak units that started firing at us, tried
to return fire but couldn't depress my gun enough because we
were right on top of it'.

Rather than skirting a white-painted church with a high steeple
near the town centre at low level, Holden opted to fly over it.
Moments later, his aircraft was hit by lines of red and green tracer
that flashed upwards, biting into his starboard wing. Almost immediately the aircraft was engulfed in flames. 'There was poor old
Holden up about four hundred feet or more, being shot to blazes
and on fire.'[9]

Holden's aircraft went out of control, diving and veering
sharply to port. Les Knight had to haul on the controls to avoid a
collision with his leader's aircraft and within seconds Holden's
Lancaster had crashed and exploded with the loss of everyone –
Guy Gibson's Dambusters crew – on board. It was Holden's
thirtieth birthday. 'They just dived down, nearly hitting us on the
way,' Fred Sutherland recalls, 'straight into the ground with a
huge explosion. It was some sight – eight guys just dying in front

of my eyes. They didn't have a hope. It was so close I could almost reach out and touch it. Your friends are getting killed and you are scared as hell but you can't let it bother you; if you did, you could never do your job. You just think, "Thank God it wasn't us."'

The blast from the crash almost brought down the two Lancasters flying close behind him, but, fortunately, his delay-fused 12,000-pounder didn't explode. Had it detonated on impact, all the other aircraft in his formation, flying as low as 30 feet above the ground, would almost certainly have been destroyed as well. Instead an unfortunate German family whose farm was the site of Holden's fatal crash suffered a tragic blow when the bomb detonated fifteen minutes later. Alerted by the anti-aircraft fire and the thunder of the approaching bombers, the farmer, his wife and their six children had been sheltering in a cellar beneath their farmhouse when the crash occurred. However, a few minutes later, the parents decided to go back upstairs to fetch some warm clothing for their shivering children. They were still above ground when the bomb exploded. The farmer survived, sheltered by one of the few pieces of wall to survive a blast that demolished every farm building and set fire to an avenue of oak trees, but his wife was killed instantly. She was the only German fatality from the raid.[10]

The remaining aircraft re-formed, with Mick Martin taking over as leader, but ran into low-lying mist and fog over the Dortmund area, at times reducing visibility to as little as 500 yards. The haze was reflecting the moonlight and 'making the whole scene appear like a silver veil. We could see practically no ground detail when flying into-moon.'[11] They were supposed to bomb from 150 feet at two-minute intervals, but the fog and haze meant that the only time they could actually spot the canal was when they were already directly overhead and too late to drop their bombs on it. They kept circling, hoping for a break or eddy in the fog that would give them a sight of the target, but with no sign of the Mosquitos that were supposed to act as 'can-openers', suppressing the air defences, the Lancasters were making themselves 'sitting ducks for the air defences putting up a wall of flak', and the prowling night-fighters. They soon lost another aircraft, when Mick

Martin's rear gunner suddenly called out, 'There goes Jerry Wilson.' Flight Lieutenant Harold 'Jerry' Wilson's Lancaster had been hit by anti-aircraft fire and he crashed into the canal bank, killing himself and all his crew.

'There was only us and Micky Martin left by then from our formation,' Les Knight's front gunner, Fred Sutherland, says. 'It was a desperate scene unfolding around us; it was pretty scary.' Soon afterwards, squinting into the fog, Sutherland froze as 'trees atop a ridge just appeared in front of me, rushing towards me. Someone else screamed for Les to climb but it was too late.' Sydney Hobday, the crew's navigator, remembered the moment all too clearly, 'To my horror,' he says, 'I saw the treetops straight ahead and thought we had at last "bought it" – after quite a good run for our money I admit!'

The trees hit them on the port side, puncturing the radiators of both port engines and damaging the tail. Both port engines overheated and had to be shut down, and the starboard inner engine then began to fail as well.

Knight fought to control his badly damaged aircraft as Edward 'Johnny' Johnson – not the member of Joe McCarthy's crew married to Gwyn, but another bomb-aimer with a similar name – jettisoned the 12,000-pound bomb, praying that the delayed-action fuse would work, because, if not, they'd be blown to pieces as it detonated. It fell away silently and they all breathed a sigh of relief. The crew also threw out their guns and ammunition to lose weight as Knight tried to nurse his battered aircraft back to England, alternately feathering the port engines to cool them and then briefly restarting them as the aircraft dropped towards stalling speed.

He called his rear gunner, Harry 'Obie' O'Brien, forward to haul on the exposed controls from the starboard rudder pedal to ease the strain on Knight's leg as he battled to hold the damaged aircraft in straight and level flight, but it was a hopeless task. With the two port engines virtually useless and the starboard ones overrevving as they strained to keep the Lancaster airborne, the aircraft was constantly being pushed to port and still losing altitude, with the glide angle increasing steadily. Fear of what was to come gripped them all. 'There was no smoke or flames,' Sutherland says, 'but we knew we didn't have long.' As they passed over Den Ham in the Nazi-occupied Netherlands, Knight realised he couldn't control the aircraft much longer and ordered his men to bale out. Looking out, Sutherland thought they were over water, but once more it was just the moonlight reflecting from the layer of cloud below them, and when he pulled back the blackout curtain he saw the ground in front of them.

The crew baled out one by one. 'Bomb-aimer going, cheers, Les,' Edward Johnson said.

'Cheers and good luck, Johnny,' Knight said, his voice showing none of the emotion he must have been feeling.

Obie O'Brien also said his farewell and baled out from the rear hatch, and was followed moments later by Sutherland, who called, 'Mid-upper gunner going out the back door, Les.' He didn't have his parachute on, but 'quickly clipped it on and just jumped out the back door'. The extra gunner, Les Woollard, on his first flight with Knight's crew, jumped at the same time, though Sutherland lost sight of him at once.

Navigator Sidney Hobday baled out of the hatch in the nose and flight engineer Bob Kellow followed a heartbeat later. He'd disconnected his intercom and so couldn't speak to Knight, but gave him a thumbs-up sign, and saw Knight's answering signal before he tumbled out of the hatch.[12]

None of Knight's crew had ever talked about being shot down. Sutherland says:

I don't think we ever talked about the possibility, or what we would do. I remember Johnson always wore special shoes whenever we flew low level so he could walk out if we came down – he was prepared. But for me, whatever was going to happen would happen. I didn't think too much about it. No one talked about it. We just hoped that the op would be over quickly, and we'd survive and get back to the Mess for a beer![13]

However, the first thing that every aircrew member found out about a new aircraft was 'how to leave the plane in a hurry'. At one time crews practised baling out from a static aircraft on the ground, but 'this produced so many twisted and broken limbs that it was put on hold'. An instructor at OTU – the Operational Training Unit, which all ranks had to attend before joining a unit on active service – also had a warning for trainee aircrew who baled out over the UK: 'Remember to hold on to the ripcord handle and bring it back or you will be charged five bob for its replacement!'

Like the rest of his crew, Sutherland had never used a parachute before, but after a heart-stopping pause when he pulled on the ripcord, his chute opened safely. 'I hit the ground and stood up,' he says. 'A few hours before I'd been in England, now I was stand-ing in enemy territory. It was quite a shock. I thought about my family getting a telegram to say I was missing, what would they think?' As he did so, he saw Les Knight attempting a forced land-ing a quarter of a mile away, but sadly, by waiting for his crew to bale out, time had run out for Knight himself, 'a classic example of the pilot sacrificing his life to allow the others to escape'.[14] His stricken aircraft hit the trees, crashed and burst into flames, killing Knight instantly. His body, still at the controls, was retrieved by Dutch civilians who, in defiance of the German occupiers, buried him after conducting a funeral for him. 'I owe my life to Les,' Sutherland says. 'He kept the aircraft steady as long as he could, so we could get out. Without him, I'd have been dead.'

Sidney Hobday, who was a Lloyds clerk in peacetime, had also landed safely – albeit 30 feet up in the branches of a tree – and saw his skipper's last moments. 'I imagine that when he let go of the stick, the plane dived straight to the deck … I shall never forget how he wished me good luck before I left … he was a good lad.'[15]

Mick Martin had lost sight of Knight's aircraft in the fog and did not know what had happened to him. He eventually identi-fied the target 'after stooging around for about an hour, but it was very hairy', and he had to make thirteen passes over it before his bomb-aimer was sufficiently confident to release their bomb.

Meanwhile, more of their comrades were being shot down. Flying Officer William Divall's Lancaster came down a few miles away after being hit by flak. Having dropped his bomb into the canal, Divall crashed into the bank and the ensuing explosion flattened the trees flanking the canal and blew the rear turret, with the rear gunner's body still inside it, right across to the opposite bank. All the crew died in the blast.

Flight Lieutenant Ralf Allsebrook's Lancaster was also hit by flak as he flew over the canal. A veteran of two tours with 49 Squadron, Allsebrook had joined 617 Squadron a few days after the Dams raid, and was not to survive his first op over Germany with them. He tried to make an emergency landing, but hit the roof of a house and then smashed into a crane on the canal bank, decapitating himself and killing his crew.

The lethal anti-aircraft fire and the crashes caused by low-level flying in such poor visibility made it unsurprising that only two bombs – dropped by Mick Martin's and Dave Shannon's crews – landed anywhere near the target, one hitting the towpath, the other falling in the water without doing any significant damage to the canal. Even worse, the abortive raid had seen five of the eight Lancasters shot down or crash, leaving a trail of burning aircraft across the German countryside, and causing the loss of forty-one men's lives, including thirteen of the survivors of the Dams raid that had made the squadron's reputation. The op had also claimed the lives of David Maltby and his crew the previous night, making a total of six out of nine aircraft and their crews lost – a loss rate of two-thirds compared with the 5 per cent losses that the supposedly more vulnerable Main Force bombers were suffering on their mass raids on heavily defended German cities.

The first two major ops by 617 Squadron had therefore cost the lives of fourteen crews. The death rate on the Dortmund–Ems Canal op was equivalent to that of the triumphantly received Dams raid, and as Johnny Johnson remarked, 'In many ways it was not dissimilar to the Dams, apart from those very heavy defences and the difficulty of getting at the target. That was the killer.'[16] Yet while the Dams raid had been hailed as one of the greatest successes of the war, the failure to destroy the target this time caused Dortmund–Ems to be regarded as an unmitigated disaster.

The margins between great success and total failure were proving to be vanishingly small.

Johnny Johnson was ill and had played no part in the raid, but hearing about the losses, he was desperate to find out if his pilot, Joe McCarthy, and the rest of his usual crew had been involved. 'It was a worrying time, these men were my family,' he says, but to his great relief he found that they had not taken part in the raid, with both McCarthy and Les Munro temporarily grounded on the orders of the Medical Officer.

The squadron's relative inactivity since the Dams raid and the attendant 'one op' gibes from other squadrons may have led to the target and the method of attacking it being hastily chosen, with too little thought about the potential pitfalls, and as successor to the now legendary Guy Gibson, George Holden may also have been eager to win his own spurs. 'There was a sense that we had to get back on ops,' Johnny Johnson now says. 'Squadron Leader Holden wanted to do it to keep the reputation of the squadron and its role in special operations. Maybe the accolades we had received because of the Dams op meant we had to get on and do more, be more successful. But those accolades were a hindrance here.'[17]

Only six crews – including those of Mick Martin, Dave Shannon, Les Munro and Joe McCarthy, who were veterans of the Dams raids – now remained on the squadron. Desperate to atone for the failures, Martin volunteered to return to the target the following night, flying solo to complete the job, but he was overruled by his superiors and, apart from an abortive attempt to bomb the Anthéor railway viaduct in southern France the following night, the Dortmund–Ems Canal raid proved to be 617 Squadron's last for almost two months.

The heavy losses they had suffered at the canal were proof that their signature operations – low-level, night-time, precision-bombing raids – were no longer viable. They had been lucky at the dams, albeit still with the loss of almost half their force. At the Dortmund–Ems Canal their luck had run out. 'It was a big blow to the squadron,' one crewman says. 'We lost so many that night that it seemed to affect the thinking of the powers that be. It was a very traumatic experience.'[18]

There were to be no more low-level attacks. From now on 617 Squadron would operate at high level, using a new tachometric precision bombsight, the SABS (Stabilised Automatic Bomb Sight), to ensure accuracy. It was one of the world's first computerised bombsights and a complex, hand-built piece of equipment, consisting of a mechanical computer mounted to the left of the bomb-aimer and a stabilised sighting head fitted with an optical graticule. The sight was connected to a Bombing Direction Indicator (BDI) mounted on the pilot's instrument panel, which indicated the amount of left or right turn required to bring the sight to bear on the target. Once the sight had been programmed with the necessary data – the aircraft's speed and altitude, and the wind-speed and direction – the bomb-aimer had only to keep the target centred in the graticule and the sight itself would then automatically release the bomb at the right moment.

However, while they could achieve impressive accuracy with the sight, and attacking from height made them less vulnerable to flak, it also made them much more vulnerable to German night-fighters, particularly when attacked from below, the Lancaster's blindspot. In the early stages of the war, anti-aircraft guns had claimed far more victims than fighters, but that was quickly reversed and by 1943, Bomber Command losses to night-fighters were twice those caused by flak.

German night-fighter pilot Peter Spoden – these days a great-grandfather living in a care home with his wife – brought down twenty-four four-engine British bombers during the war, and he cries as he reflects on the deaths of the crews inside them, young men of his own age. The aircraft he shot down never even knew he was there: he approached from behind and below them, flew 50 or 60 feet underneath their fuselage and unleashed the two upward-firing guns the German pilots called *Schräge Musik*. (Translating literally as 'slanting music', *Schräge Musik* was their slang term for jazz.) Spoden recalls one night where he was talked in by his radio operator and suddenly saw 'this black shadow above me ...

in ten minutes I shot down three Lancasters – I was completely out of my mind.'[19] However, the firing wasn't all one way. One Lancaster gunner has vivid memories of shooting down a fighter at close quarters: 'I could see my bullets hitting him. I couldn't miss him – not at that range.'[20]

617 Squadron's shocking rate of losses had led to their sarcastic nickname 'The One Op Squadron' being replaced with a new one, 'The Suicide Squadron', and the deaths of so many crewmates dealt what could easily have been a terminal blow to morale. 'Those losses had a big effect, there was a sense of distress and shock, and possibly even dissatisfaction that we were asked to do something which should never have been attempted,' Johnny Johnson says. But although morale was inevitably affected in the short term, confidence soon recovered. 'Morale slumped because they were rather staggering losses,' Larry Curtis adds, 'but one did tend to throw these things off very quickly. Going from low level to high level made all the difference; losses were very slight after that.'[21]

★   ★   ★

While their comrades were trying to come to terms with the disaster, two of the survivors of Les Knight's crash had been captured, but the remaining five, including Sidney Hobday and Fred Sutherland, were on the ground in the Occupied Netherlands, trying to evade the Nazis. They were separated from each other, and the knowledge that he was now alone in the heart of enemy territory, facing capture or perhaps even death if he were found, almost paralysed Hobday at first. However, realising that the greatest danger of discovery lay in remaining close to the wreckage of his downed aircraft, he climbed down from the tree he had landed in and set off south, away from the burning Lancaster.

He walked through dew-soaked meadows and along a canal bank, carrying on until it started to get light, when he hid in a small wood. However, his feet were soaked, and, sitting on the wet grass, he began to feel very cold. 'Not wishing to get pneumonia,' he began walking again, but as he approached a metalled road, the sound of galloping hoofs terrified him and he dived behind the nearest hedgerow, imagining 'a couple of dozen mounted Jerries looking for me'. When he risked peering out, he saw that the 'hoof-beats' were actually the noise made by some Dutch children's wooden clogs as they ran along the road to school. As he waited for them to pass, he glanced at his watch. It was eight-thirty in the morning. 'Twelve hours before, I had been strumming the piano in the Mess.'[22]

Before setting out along the road, he took off his brevet and his other RAF markings, trying to make his battledress look as civilian as possible. Hobday knew that his name and those of his comrades decorated after the Dams raid had been published in the English newspapers, and as a result they had all been put on a Nazi blacklist. He knew that if he was taken prisoner, he was unlikely to remain alive for long.

He had not walked far when he saw two farmworkers cycling towards him. He bent down, pretending to tie his laces, but they stopped. Not speaking Dutch, he couldn't understand them, but after a few moments of gut-gnawing indecision, he decided to risk telling them who we was. He said 'RAF' several times without any sign of recognition from them, and then began flapping his arms about to mimic flying. They now seemed to understand and, having looked carefully up and down the road, gave him half their food, 'black bread with some queer stuff in it which I could not stomach'. He gave them a couple of cigarettes in return from the packs he always carried on ops, in case of just such an eventuality.

Heartened by their friendliness and realising the impossibility of crossing Europe alone and unaided, Hobday decided to seek more help from civilians where he could, hoping they would put him in touch with the Dutch Resistance. After a few more hours of walking he tried to hitch a ride in a little cart, but the driver shook his head, indicating by sign language that the Germans would slit his throat if they caught him. However, he gave Hobday some more black bread before driving off.

A little further down the road, he saw the same cart driver in urgent conversation with a woman, who then passed Hobday on her bicycle a couple of times, studying him carefully without speaking. Once more he was left fearing betrayal to the Nazis, but he kept walking and was then overtaken by some young men, who spoke to him in 'slow schoolboy English'. They gave him some apples and a tall man then brought him a civilian suit. It would have 'fitted a man five inches taller than myself,' Hobday said, but he changed into it and the Dutchmen took his RAF uniform away. They also insisted on shaving off his moustache, saying it made him look 'too English'.

Hobday was then told to make his way alone to a railway station 10 miles away, as it was too dangerous for them to accompany him. By the time he arrived, he was close to exhaustion. He hadn't slept for thirty hours and had walked for another twelve with almost no rest. The tall man was waiting for him and gave him a train ticket to a town 100 kilometres away with a list of the times of the trains he had to catch. He also gave him a note in Dutch that said: 'This man is deaf and dumb. Please help him.'

The journey tested Hobday's nerves to breaking point. He first almost blundered into a carriage reserved for Wehrmacht troops and then, when he found an empty carriage, a 'German Luftwaffe man and his girl' got in and sat next to him. Luckily they were more interested in each other than the strange man sharing the compartment, and with the aid of his 'deaf and dumb' note, Hobday made it safely to his destination, where a young member of the Dutch Resistance met him. Having questioned Hobday searchingly to make sure he was not a German spy, he led him out of town to a place where eight members of the Resistance were in hiding, living in a crude hut deep in the heart of dense woodland. They had been carrying out minor acts of sabotage and raiding German stores, assembling 'quite a collection' of firearms, explosives, uniforms, blank visas and identity cards.

Twenty-four hours later, Hobday was reunited with Fred Sutherland, who had also managed to make contact with the Resistance. Fred had walked a few miles from the Lancaster's crash site when, realising that 'walking all the way to the south of Europe was never going to work', he hid behind a barn and then jumped out as a girl about his own age was cycling towards him.

'She nearly jumped out of her skin!' he says. 'She couldn't speak any English so I tried to communicate with sign language that the Germans would cut my throat if they caught me.' She took him to a boy who could speak a few words of English, and he contacted the Resistance. 'After the war, I was told that this girl had actually been dating a German soldier!' Sutherland says. 'So I guess I was lucky because she didn't tell anyone.'

Fred Sutherland

Sutherland and Hobday were comfortable enough living in the hut, sleeping on stolen German blankets and straw beds. Their food was largely potatoes, although one day a Dutchman caught some tiny eels in a nearby canal. The Resistance had begun making arrangements for the two RAF men to be returned to England via France and Spain, but the long chain of helpers was vulnerable to infiltration or arrest by the Nazis, and it proved a lengthy and fear-ridden process. Twice they were almost discovered, once when German troops began holding infantry manoeuvres in the woods, and the other when they escaped a Gestapo raid on the hut by the skin of their teeth.

After three weeks, frantic to contact his wife, who he knew would believe that he had been killed, Hobday had to be prevented from setting off for Spain on his own, but a week later arrangements were finally in place. The night before their departure, their hosts staged a farewell party for them, fuelled by a bottle of gin and some beer. The next day they set off, first travelling to Rotterdam, escorted by a woman dressed as a nurse.

They then travelled to Paris by train, armed with new fake identity papers showing that they were labourers for the Todt Organisation working on an aerodrome near Marseille. (As the Third Reich's Minister for Armaments and Munitions, Fritz Todt ran the entire German construction industry. His *Organisation Todt* built the West Wall that guarded the coast of German-occupied Europe, as well as roads and other large-scale engineering projects in occupied Europe.) They went via Brussels and had 'some shaky moments' at the two frontiers, surviving a close examination of their fake identity papers at a German checkpoint. When the German officer held them up to the light for a better look, Sutherland's hands were shaking so much that he had to ball his fists and brace his elbows against his side to hide them. 'My heart was pounding and I was really scared,' he said. 'It's hard to be nonchalant when you are facing your enemy.' However, with the help of their Dutch escorts, who, at considerable risk to their own lives, kept up a stream of distracting conversation with the German frontier guards, the fake papers passed scrutiny.

'I can't begin to describe the courage of the people who helped us in Holland and France,' Sutherland says. 'They took us into their homes, fed us and cared for us at tremendous risk to themselves and their families. The Germans had infiltrated the Underground and people did not know who they could trust, and yet still they helped us, even knowing that, while we would likely be sent to a prisoner-of-war camp, they and their families would be shot.'

They remained in Paris for nearly a fortnight, staying in the tiny flat of an elderly French lady, at huge risk to herself, and eventually they were taken to a clearing house for escaping aircrew and PoWs. There they were given yet more new papers and then set off in small groups for the journey to the Pyrenees.

When they arrived at Pau, they got themselves French-style berets and then took a small train through 'the most beautiful scenery I have ever seen' to Sainte-Marie and were driven on from there in a car powered by gas made from charcoal. At the foot of the Pyrenees they lodged overnight in a barn where other escapers were already waiting and began the climb of the mountains the next day. Apart from their guide and his dog, there were ten escapers: three Americans, three Frenchmen, a Dutchman, an Australian, the Canadian Sutherland and Hobday, the only Briton.

Following weeks in hiding, on a very poor diet and with little chance of exercise, Hobday was very unfit. Even worse, after climbing for six hours over the rocky paths, his shoes fell apart. Fortunately the guide had a spare pair, though they were too small and 'hurt like hell'. They climbed all night, a perilous ascent with no light to guide them, following narrow, twisting paths with the mountainside rising sheer above them on one side and a sheer drop on the other. Only the thought of the fate that awaited them if they were found by the Nazis spurred them on. They had little rest and even less food, and suffered a frustrating and frightening delay when a shepherd, recruited by the guide to show them a short cut to the Spanish side of the mountains, became completely lost and left them in driving rain 7,000 feet up on the mountainside, while he tried to discover where they were.

They had started climbing the mountains at seven o'clock on Wednesday evening and did not reach the Spanish side until the Saturday morning. Having already passed through the Netherlands and right across France, in constant fear of discovery by the Gestapo, they had then dragged themselves right over the Pyrenees. Their epic escape was 'the toughest thing I've ever done', Sutherland says. Completely spent, they rested for the remainder of that morning and swallowed some food and wine, though it 'came up as fast as it went down'.

In the afternoon they walked down to the nearest village, Orbaizeta. By then Hobday was so stiff he could hardly walk, and his companions were little better. Although Spain was ruled by Franco's fascist regime, it was professedly neutral in the war, but there was a tense atmosphere as they encountered the Spanish carabineros for the first time. However, they treated the escapers well enough, and they remained in the 'dirty little village' until the Monday, though Hobday had to sell his watch to pay for food for Sutherland and himself. The shop where they ate was 'a general store, very much like the Wild West saloons of the old cowboy films, complete with liquor, shepherds, singing and a bit of good-natured scrapping. On the Sunday they all came in with their week's money and proceeded to get rid of it on booze.' The place was filthy and there were pigs and chickens wandering everywhere, indoors and out.

The escapees were then taken to Pamplona, where they were met by the Red Cross, who escorted them to Madrid, a journey that took a further fortnight. There staff at the British Embassy gave them a train ticket to Gibraltar, where, to their enormous relief, they were at last back on British soil. Hobday's first action was to cable his wife to tell her he was alive. They were flown home a few days later, on 6 December 1943, almost three months after they had been shot down.

If Hobday needed any reminder of how fortunate they had been to come through that marathon journey unscathed, the fate of a Dutchman he had befriended provided it. He attempted to cross the Pyrenees a week after Hobday but was caught in a snowstorm and got lost. Suffering from frostbite, he was captured by a German frontier guard and sent to Buchenwald concentration camp.[23]

Fred Sutherland, speaking from his home in Canada seventy years on, perhaps encapsulates the emotions of all those wartime evaders:

> *The whole experience was quite unreal, just like living in a movie. It was all very nerve-racking and I didn't rate my chances of making it home, but it was worth a try. The Dutch people were risking their own lives to help me. Without them I would have been captured or dead and I can never thank them enough for that. And when I was on the ground – seeing the Nazis close up – that's when I realised we had to win this war; regardless of the cost.*

★   ★   ★

For the families of the dead, the ramifications of the disastrous Dortmund–Ems raid went on for months as they struggled to come to terms with their loss, and to understand how their loved ones had perished. David Maltby's father, Ettrick, received a letter from the mother of Maltby's navigator, Vivian Nicholson, desperate to find out how her own son had died. She was heartbreakingly eloquent in her expression of her grief.

'I scarcely know how to write this letter,' Elizabeth Nicholson wrote.

> *We would like to know the true facts of what they did that night and would be gratefully thankful for any news you could give us. We have a photo of your gallant son and our boy together. It is indeed a terrible wound for us to see them so young, happy and beautiful. Our boy was conscientious, very guarded about his duties. Please, if your son told you anything [about our son], we would indeed be grateful if you could let us know. I know you will understand my yearning for news, and I have the worry of our second son aged 19 on the submarine HMS Seanymph. The world owes so much to these gallant young men. We can only wait patiently till we can understand why they are taken from our homes, where their places can never be filled.* [24]

Many of the young men taking to the skies that night had left 'last letters' to be delivered to their loved ones if they were killed. Maltby's wireless operator, Antony Stone, was no exception. His mother received her son's final letter shortly after his death. 'I will have ended happily,' he wrote, 'so have no fears of how I ended as I have the finest crowd of fellows with me, and if Skipper goes I will be glad to go with him.' [25]

Many more letters, to parents of the bereaved, and from those who had made the ultimate sacrifice, would be delivered before the war was over.

CHAPTER 3

Press On, Regardless

Bomber Command's Main Force had continued to pound Germany during the autumn and early winter of 1943, including a raid on Kassel on 22/23 October that caused a week-long fire-storm in the town and killed 10,000 people, and the launch of the 'Battle of Berlin', a six-month bombing campaign against the German capital, beginning on 18/19 November. Over the course of the campaign some 9,000 sorties would be flown and almost 30,000 tons of bombs dropped. Bomber Harris's claim that though 'it will cost us between 400–500 aircraft. It will cost Germany the war'[1] proved optimistic; the aircraft losses were over twice as heavy and the bombing campaign did not destroy the city, nor the morale of its population, nor the Nazis' willingness to continue the war. In contrast, the heavy losses sustained by Main Force during this period undoubtedly had an effect on morale.

Following George Holden's death at Dortmund–Ems, Mick Martin had been put in temporary charge of 617 Squadron, but because he had jumped two ranks from Flight Lieutenant direct to Wing Commander, some felt his face still did not fit, and on 10 November 1943 he was replaced by a new commander, a tall, dark-haired and gaunt-faced figure: Wing Commander Leonard Cheshire. Still only twenty-six, Cheshire was one of the RAF's most decorated pilots and its youngest Group Captain, who had taken a drop in rank in order to command 617.

Cheshire made a dramatic entrance to his new squadron. Gunner Chan Chandler was in his room at the Petwood Hotel when there was a shot directly outside his window. 'I looked out and saw this chap with a revolver in his hand, called him a bloody lunatic and asked what the hell he thought he was doing. He replied that he thought the place needed waking up! It turned out he was the new CO and my first words to him were to call him a bloody lunatic! However, we got on famously after that.'[2]

Leonard Cheshire

At his first squadron meeting Cheshire told his men that 'If you get into trouble when you're off duty, I will do what I can to help you. If you get into trouble on duty, I'll make life a hell of a lot worse for you.'

'So we all knew where we stood from the start,' Johnny Johnson says. Far more approachable and friendly than Guy Gibson, Cheshire was soon highly thought of by all. 'He developed the techniques for marking, for instance,' Johnson says, 'so the efficiency of the squadron was improved. That was always his objective: to get things absolutely right.'

Larry Curtis found Cheshire 'very quiet, very relaxed, but at the same time, you did what you were told; he just approached it in a different way.'[3] Malcolm 'Mac' Hamilton witnessed this new approach when he joined the squadron. His crew had achieved an impressive bombing accuracy of 70 yards before joining 617, and when he met Cheshire for the first time, Cheshire said, 'Oh, Hamilton, I see your crew have won the 5 Group bombing trophy three months running.'

Hamilton acknowledged this, 'putting my chest out, because I was quite proud'.

Cheshire frowned. 'Well, I tell you what I'll do. I'll give you six weeks to get your accuracy down to twenty-seven yards, or you're off the squadron.'[4]

He was only half-joking. On 617 Squadron, everything revolved around accuracy in bombing. Before they were allowed on ops, every new crew was required to carry out at least three six-bomb practice drops on the ranges at Wainfleet on the Wash in Lincolnshire. The crew's accuracy was assessed and the results added to the Bombing Error Ladder kept by the Bombing Leader. Crews with the highest accuracy – and the results of even the most experienced crews were continually updated – were assigned the most ops and the most important roles on those ops, and those with poorer records would find themselves left out, or even transferred off the squadron altogether, if their results did not improve. Cheshire was relentless in raising standards, but he set himself the highest standards of all, and was universally admired and even loved by the men under his command. 'He was a great man,' Johnny Johnson says, 'and the finest commander I served under.'[5]

★   ★   ★

In the autumn of 1943, during the continuing lull in ops that followed the disaster of Dortmund–Ems, 617 had been practising high-level bombing with new 12,000-pound 'Blockbuster' HC bombs – the equivalent of three 4,000-pound 'cookies' bolted together.

One of the squadron's rear gunners, Tom Simpson, was lugging his Browning machine guns to the firing range one morning with 400 rounds of ammunition draped around his neck, when a 'silver-haired, mature gentleman' approached. A civilian on the base was an unusual sight, but the man seemed to know his way around. He asked if he could walk with Simpson and helped to carry one of the guns. After Simpson had installed the guns in the practice turret and fired off a few bursts himself, he noticed the stranger's 'deep blue eyes gleaming' and asked, 'Would you like to sit in here and have a little dab yourself?'

The man didn't need a second invitation. Having watched him fire both guns singly, Simpson invited him to fire off the last hundred rounds using both guns together. After doing so, he was 'trembling with excitement and pleasure' as he climbed out of the turret. 'That was really fantastic,' he said. 'I had no idea of the magnitude and firepower a rear gunner has at his disposal.' Only later did Simpson realise that the silver-haired, mature gentleman was Barnes Wallis.[6] When he mentioned the incident to Mick Martin, he told him that Wallis was developing even bigger and better bombs for the squadron. Within a few months they would all have the proof of that.

Barnes Wallis

Wallis, whose wife's sister and brother-in-law had been killed by a German bomb in 1940, had already devised Upkeep – the 'bouncing bomb' used on the Dams raid. Forever innovative and forward-looking, he had never believed that carpet bombing could break German resolve, any more than the Blitz had broken Britain's will to fight. Instead, he had argued from the start for precision bombing of high-value economic, military and infrastructure targets, and designed a series of weapons capable of doing so. He had been given the task of creating newer, ever more destructive weapons that could penetrate and destroy heavily fortified concrete bunkers and other targets, previously invulnerable to conventional attack.

A vegetarian non-smoker, Barnes Wallis came from a respectable middle-class background – his father was a doctor and his grandfather a priest – yet at sixteen years old, he had set his face against the advice of his parents and teachers and left school to take up an apprenticeship as an engineer. In later years, his daughter Mary attributed that decision to an experience when he was very young and his mother had taken him to a foundry to see the men and machinery at work. 'The size, power, the noise of machinery in the light of the flames from the foundry furnace' may have made a lasting impression on him.

Whatever the reasons, Wallis's career path was set, though at first he showed more interest in the sea than in the air, training as a marine engineer, before being recruited by Vickers to work on airship design. An engineering genius, he worked as the chief designer on the *R100* airship, 'a perfect silver fish gliding through the air … a luxury liner compared with the sardine-tin passenger aircraft of today'. Having pioneered the geodetic system of construction in airships (more commonly called geodesic) – a latticework system of construction that produced a very light metal structure that nonetheless possessed great strength – he went on to apply it to military aircraft too, first using it on the airframe of the Wellington bomber. Despite his considerable achievements, he was a warm, humane and profoundly modest man. 'We technical men like to keep in the background,'[7] he said, and a friend and work colleague remarked that: 'He would sooner talk about his garden than himself.'

Although he had designed aircraft, Wallis knew very little about bomb construction, but right from the outset of the war he resolved to put that right as quickly as possible. One of his early discoveries was that the explosive power of a bomb is proportional to the cube of the weight of the charge it carries, so that, for example, a 2,000-pound bomb would have eight times the impact of a bomb half its weight. He also learned that the pressure wave from an explosion is transmitted far more efficiently through the ground or through water than it is through air. Both of those discoveries were reflected in the design of the 'Upkeep', 'Tallboy' and 'Grand Slam' bombs he subsequently produced.

In addition to the new bombs Barnes Wallis was producing, 617 Squadron were now using a new and far more accurate bombsight. Shortly before Leonard Cheshire's arrival, 617 had begun training with the sophisticated new SABS. Its only major drawback was that it required a long, straight and level run to the dropping point, reducing the pilots' ability to make evasive manoeuvres and increasing their vulnerability to flak and fighter attack.

Wireless operator Larry Curtis always dreaded that vulnerable time, flying straight and level with the bomb-aimer in virtual control of the aircraft:

> *As long as you were busy, you never thought about anything except the job you were doing. But when it came to the bombing run, and to a great extent your duties were finished, it was then that you became aware that you were very vulnerable and people were actually trying to kill you. In the radio operator's cabin there was a steel pole – some sort of support – and I used to hang on to it like grim death. I've always said if anyone could find that steel pole, they'd find my fingerprints embedded in it – and I'm not joking. It did come home very forcefully; I'm sure everyone would agree that was the time you dreaded.*[8]

There was no man on the squadron who did not feel fear at times, but, says one of the squadron's wireless operators, 'it was all about pushing fears to one side. I think if anyone said they were never afraid, I'd say they were either not there, or they're lying. There was no possibility you couldn't be scared at some points.'[9]

Even the squadron's greatest heroes were not immune to fear. Wilfred Bickley, an air gunner who joined 617 at the same time as the great Leonard Cheshire, VC, once asked him, 'Do you ever get scared?'

'Of course I get scared,' Cheshire said.

Bickley broke into a broad grin. 'Thanks for that, you've made my day.'[10]

At a reunion after the war, Cheshire also said to one of his former men on 617 Squadron: 'I could have been a pilot or at a pinch a navigator, but could never have done the other crew jobs; there was too much time to think, to be isolated, to dwell on what was going on around you. I never felt fear as a pilot, but as a passenger, I certainly felt it!'[11]

★　　　★　　　★

The use of the new SABS bombsights in training on the ranges at Wainfleet had led to a huge improvement in accuracy, but the first real chance to assess their effectiveness under operational conditions came in raids against Hitler's new 'V-weapons'. 'V' stood for *Vergeltungswaffen* ('vengeance weapons'), and these 'terror weapons' were the Nazis' chosen method of retaliation for the relentless Allied bombing of German cities. Lacking the heavy bombers and air superiority required to inflict similar damage on Britain, Hitler had authorised the development of three V-weapons: the V-1 'flying bombs' – an early forerunner of modern-day Cruise missiles, christened 'buzz-bombs' or 'doodlebugs' by Londoners – the V-2 *Feuerteufel* ('Fire Devil') rocket, and the V-3 'Supergun'. Reinforced underground sites were being constructed in the Pas de Calais, in occupied France, where the weapons could be assembled and then launched against Britain. Most of the work was carried out by the forced labour of concentration camp inmates, PoWs, Germans rejected for military service and men and women from the Nazi-occupied countries. They were slave labourers, worked round the clock and fed near-starvation rations, and many did not survive.

Destruction of the plants where the V-1s were being manufactured and the ramps from which they were to be launched had become an increasingly urgent priority. 617's aircrews were briefed about the sites and what they meant. 'We were trying to obliterate them before they were even made.'

Intelligence on the German V-weapon sites came both from reports by French Resistance members of unusual building activity, and from RAF large-scale photographic reconnaissance missions, which covered the entire northern French coast for up to 50 kilometres inland. Sporadic attacks were launched by Main Force squadrons but inflicted only superficial damage.

The task of finding and eliminating the V-weapon sites was made harder by the camouflage used by the Germans – some nets covering buildings were 'painted to look like roads and small buildings'.[12] The wooded terrain in which most of the weapons were sited also hindered attacks, but the biggest problem was the poor visibility and persistent fogs that shrouded the Channel coast of Occupied France for many winter weeks.

On the night of 16/17 December, 617 Squadron joined the assault on Hitler's vengeance weapons, with a raid on a site at Flixecourt, south of Abbeville, where they were led into battle for the first time by Leonard Cheshire. A system of markers was use to aid accuracy in navigation and bombing. Parachute markers that floated slowly down were used to mark both turning points on the route to the target and the target itself, and target indicators, burning in different colours, were also utilised. On this op, a Mosquito from Pathfinder Force – the specialist squadrons that carried out target marking for Main Force ops – was marking the target from high level using Oboe (the ground-controlled, blind-bombing system that directed a narrow radio beam towards the target). However, the target indicators it dropped were 350 yards from the centre of the target, and such was 617 Squadron's accuracy with their new SABS – their average error was less than 100 yards – that, although they peppered the markers with their 12,000-pound bombs, none of them hit the target.

Although unsuccessful, the failure in target marking at Flixecourt did have positive consequences, for it served as the catalyst for Cheshire and his 617 Squadron crews to begin lobbying for a change in the marking system. Deeply frustrated by the failure of the raid, despite the phenomenal accuracy of their bombing, Mick Martin argued vehemently that it was pointless for the squadron to be dependent on markers dropped from height when flying missions that risked, and often cost, the lives of 617's aircrews. Far better, he said, for himself or somebody else in the squadron to take on the responsibility of marking the target at low level, but at first 'Nobody seemed to listen,' one of his crewmates recalled, 'and I think Mick got sick of volunteering and making his suggestions.'[13]

However, Leonard Cheshire may already have been thinking along similar lines, and he became an equally fervent advocate of low-level marking. The 5 Group commander, Air Vice Marshal Ralph Cochrane, lent his support to the idea, driven in part at least by his intense personal rivalry with Air Vice Marshal Don Bennett, the commander of 8 Group, which included the Pathfinder Force. Cochrane lobbied hard for a trial of the low-level system and also argued forcibly that 'his' 617 Squadron could mark and destroy targets that were beyond the capabilities of 8 Group and its Pathfinders. Cochrane's lobbying eventually proved successful, but if it was to be anything more than a short-term experiment, 617 would have to back up his words with actions.

Attempts to follow up the Flixecourt raid with a series of further attacks on V-weapon sites that December were aborted because of bad weather and poor visibility, but such conditions were, of course, no deterrent to the construction of the unmanned V-1 flying bombs that would eventually strike London.

The squadron strength for the ongoing war against the V-weapon sites was boosted in January 1944 by the arrival of several new crews. Among them was one lead by an Irish-American pilot, christened Hubert Knilans but always known as Nick, and so inevitably nicknamed 'Nicky Nylons' by his crewmates (nylon stockings, obtainable only across the Atlantic, were a rare commodity during the war, and a welcome gift for women); his good looks and American accent made him a magnet for the English girls. He came from a farming family in Wisconsin, but in 1941 was drafted for US military service. He wanted to be a pilot, not a soldier, but knew that the USAAF required all pilots to have a college degree, so, without telling his parents, he packed a small bag and set off for Canada. He arrived there literally penniless, having spent his last ten cents on a bus ticket from Detroit to the border, but he was following such a well-trodden path that the Canadian immigration officer merely greeted him with 'I suppose you've come to join the Air Force?' and directed him to the RCAF recruiting office, where he signed on to train as a pilot.[14]

He soon developed a taste for the practical jokes that all air-crew seemed to share. In the depths of the bitter Canadian winter, Knilans and a friend would slip out of their barracks at night, sneak up on their comrades pacing up and down on guard duty and, at risk of being shot by trigger-happy or nervous ones, they then let fly with snowballs. The sudden shock caused some of the more nervous to drop their rifles in the snow, and one even collapsed as he whirled round to face his attacker.

After completing his flying training, he sailed for England with thousands of other recruits on board the liner *Queen Elizabeth*. So eager was Britain to receive these volunteers that the medical and other checks they went through were often rudimentary. One New Zealander boarding his ship passed 'a friendly chap at the gangway asking him how he was as he passed by'.[15] The 'friendly chap' turned out to be the medical officer giving each man board-ing the ship his final medical examination!

After a bomber conversion course, Knilans joined 619 Squadron at Woodhall Spa in June 1943. He flew his first operational mission on the night of 24 July 1943 as Bomber Command launched the Battle of Hamburg, including the first-ever use of 'Window' – bundles of thin strips of aluminium foil now called 'chaff'. Window was dropped in flight to disrupt enemy radar by reflecting the signal and turning their screens to blizzards of 'snow'.

Informed in October 1943 that he was to be transferred to the USAAF, Knilans refused, insisting on remaining with 619 to complete his tour. His eagerness to fly almost cost him his life later that month when his aircraft was targeted by a night-fighter during a raid on Kassel. A stream of tracer shattered the mid-upper turret, temporarily blinding the gunner with shards of Per-spex, and another burst fatally wounded the rear gunner, leaving the Lancaster defenceless. Both wings were also hit, but Knilans threw the aircraft into a vicious 'corkscrew' that shook off the fighter. It was a heart-stopping, gut-wrenching manoeuvre, especially for rear gunners who, facing backwards, were thrown upwards as if on an out-of-control rollercoaster as the Lancaster dived and then plunged back down as the pilot hauled on the controls to climb again.

Despite the damage to his aircraft, the loss of his gunners, and damage to one engine, Knilans insisted on pressing on to bomb the target, before sending some of his crew back to help the wounded gunner. 'I knew that in an infantry attack, you could not stop to help a fallen comrade,' he said. 'You had to complete your charge first. Bomber Command called it "Press on, regardless".'

Despite damage to his undercarriage, including a flat tyre on the port side, Knilans eventually made a safe landing back at Woodhall Spa, using brakes, throttles, rudders and stick as he battled to keep the aircraft on the runway. He then helped the ambulance crew to remove the blood-soaked body of his rear gunner from the rear turret. He had almost been cut in half by the cannon shells that killed him. The squadron doctor issued Knilans with two heavy-duty sleeping pills so that his sleep would not be disturbed by those horrific memories, but he gave them instead to the WAAF transport driver, a good friend of the dead gunner, who was overcome with grief and unable to drive.

Awarded the DSO for his courage, Knilans wore that medal ribbon on his uniform, but not the other British, Canadian and American medals to which he was entitled. 'I thought it would antagonise others on the same squadron,' he said, 'or confirm their prejudices about bragging Yanks.'

Although he had pressed on to the target on that occasion, on another, while flying through a 'box barrage' from heavy anti-aircraft gun batteries over Berlin, with flak bursting all around them and fragments from near-misses rattling against the fuselage, his bomb-aimer told him, 'Sorry Skip, the flare's dropped into the clouds. We'll have to go round again.'

'You can still see the lousy flare,' Knilans angrily replied. 'Now drop the bombs!' The bomb-aimer got the message, 'saw' the flare and dropped the bombs, and they then 'departed in haste'.

Knilans was a popular figure on his squadron, even though he had made a deliberate decision not to become too closely involved with his crewmates. 'I would have liked to have met their families, but I decided against it. If the crew members became too close to me, it would interfere with my life or death decisions concerning them. A kind welcome by their families would add to my mental burden. It could lead to my crew thinking of me as unfriendly, but it could lead to their lives being saved too.'

By January 1944, Knilans had decided:

I did not want to go on bombing civilian populations. There were few front-line soldiers and the flak-battery operators were women, young boys and old men. The cities were filled with workers, their wives and children … This type of bombing had weakened my reliance on my original ideal of restoring happiness to the children of Europe. Here I was going out and killing many of them on each bombing raid … It was an evil deed to drop bombs on them, I thought. It was a necessary evil, though, to overcome the greater evil of Nazism and Fascism.

However, keen to remain on ops, he volunteered for 617 Squadron, which instead bombed 'only single factories, submarine pens and other military targets', and persuaded his crew to join with him. Two other squadron leaders promptly put the wind up some of them by telling them that 617 was 'a low-level flying suicide squadron', but Knilans merely shrugged and told them that he wasn't concerned about that, since 'nobody had managed to live long enough to finish a tour' on their present squadron either.

The RAF regarded a loss of around 5 per cent of the aircrews on each combat trip as acceptable and sustainable. After a quick mental calculation, Nick Knilans had 'figured it out that by the end of a first tour of thirty trips, it would mean 150 per cent of the crews would be lost. Suicide or slow death, it did not make much difference at that stage of the war.'

Two other pilots and their crews from 619 Squadron, Bob Knights and Mac Hamilton, followed his lead, telling him that 'they did not like bombing cities indiscriminately either'. It was a

short move over to 617 Squadron, as it was in the process of trans-
ferring to Woodhall Spa, ensuring that Knilans could continue to
enjoy the comforts of the Petwood Hotel Officers' Mess: 'the best
damn foxhole I would ever find for shelter'.

Knilans and his crew were allocated Lancaster R-Roger as their
regular aircraft while on 617 Squadron. The ground crew thought
it should be called 'The Jolly Roger' and wanted to paint a scantily dressed pirate girl 'wearing a skull and cross bones on her hat',
but Knilans refused. 'I did not want a scantily clad girl or a humor-
ous name painted on the aircraft assigned to me. This flying into
combat night after night, to me, was not very funny. It was a
cold-blooded battle to kill or be killed.'

Knilans did not lack a sense of humour, however, and 'carried
away one day with the exhilaration of flying at treetop level
at 200 mph', he could not resist buzzing the Petwood Hotel. 'We
roared over the roof two feet above the tiles. It must have shook
from end to end.' It was teatime and a WAAF was just carrying a
tray of tea and cakes to the Station Commander's table. 'The sudden
thunderous roar and rattle caused her to throw the tray into the air.
It crashed beside the Group Captain, my Commanding Officer. He
was not amused. Wingco Cheshire told me later that he had quite
a time keeping me from being court-martialled by my irate CO.'

★   ★   ★

On Boxing Day 1943 the last great battle of the sea war ended
with the sinking of the German battleship *Scharnhorst*, but in
terms of the final outcome of the war, an event had taken place
a month earlier that, though known to only a handful of people
at the time, was to prove far more decisive. At the Tehran
Conference of 28 November, Roosevelt and Churchill had at last
agreed to meet Stalin's constant demands that a second front
should be opened in the land war against Germany, and the invasion of France, code-named Operation Overlord, was set to
begin in six months' time, in June 1944.

617 Squadron's transfer to their new home, Woodhall Spa,
took place in early January 1944. A former out-station of RAF

Coningsby, it was now a permanent base in its own right for 617's crews and their thirty-four Lancasters. If the new base's prefab buildings were less solid than the brick-built facilities at Scampton and Coningsby, their new airfield at least had three concrete runways and thirty-six heavy-bomber hard-standings, with three hangars, a bomb dump on the northern edge of the airfield and the control tower on the south-eastern side. The roads crossing the Lincolnshire flatlands around the base were all but deserted – petrol was strictly rationed and few had any to spare – but the skies overhead were always busy with aircraft, black as rooks against the sky, though, unlike rooks, the aircraft usually left their roosts at sunset and returned to them at dawn.

Officers based at Woodhall Spa were billeted in some style in the Petwood Hotel, originally a furniture magnate's mansion and built in a half-timbered mock-Tudor style with a massive oak front door, windows with leaded lights and acres of oak panelling. The grounds included majestic elms, rhododendron-lined avenues, manicured lawns, sunken gardens and a magnificent lily pond. There was also an outdoor swimming pool, tennis courts, a golf course, and even a cinema – The Kinema in the Woods, or 'The Flicks in the Sticks' as it was christened by 617's irreverent crews – in a converted sports pavilion on the far side of the Petwood's grounds. The most highly prized – and highly priced – seats were the front six rows, where you sat in deckchairs instead of conventional cinema seats. The Petwood's beautiful grounds and timeless feel could almost have made the aircrews forget the war altogether, had ugly reality not intruded so often. As Nick Knilans remarked, 'One day I would be strolling about in this idyllic setting with a friend and the next day he would be dead.'

Inside the hotel there was a high-ceilinged, wood-panelled room adapted for use as a bar, a billiard room and two lounge areas with roaring log fires, though other parts of the building were sealed off and the most valuable paintings and furniture removed – a wise precaution given the boisterous nature of most off-duty aircrews' recreations.

The Petwood provided the aircrew with a comfortable base to escape the rigours of the war – a luxury the men of Bomber Command held dear. On a visit to one bomber base, the great American correspondent Martha Gellhorn described their preparations for an approaching operation:

> *A few talked, their voices rarely rising above a murmur, but most remained silent, withdrawing into themselves, writing letters, reading pulp novels, or staring into space. Though they were probably reading detective stories or any of the much-used third-rate books that are in their library, they seem to be studying. Because if you read hard enough, you can get away from yourself and everyone else and from thinking about the night ahead.*[16]

Cheshire himself almost became the squadron's first casualty at Woodhall Spa while carrying out an air test on 13 January 1944. Just after take-off, he flew into a dense flock of plovers and hit several of the birds. One smashed through the cockpit windscreen, narrowly missing Cheshire, while another struck and injured the flight engineer who was the only other person on board. Cheshire made a low-level circuit and managed to land his damaged aircraft safely. 617 Squadron mythology claims that at least twenty plovers were on the menu at the Petwood Hotel that night![17]

The thin Perspex of the canopy and the bomb-aimer's 'fish-bowl' in the nose were very vulnerable. 'The last thing you want as you are tearing down the runway at a hundred-plus and about to lift off is a flaming bird exploding through the canopy; for one thing it makes a hell of a bang, and sudden loud noises are not popular in an aircraft, especially in the middle of a take-off.' If they did hit a bird, the bomb-aimer and the flight engineer were 'liable to get a faceful of jagged bits of Perspex and a filthy mess of blood, guts and feathers to clear up', said Gunner Chan Chandler.[18] Bird-strikes were a serious problem and at Scampton, Coningsby and Woodhall Spa there were scarecrows, bird-scarers and regular shooting parties with half a dozen shotgun-toting aircrew touring the perimeter of the base in a van and shooting every bird they saw. 'There were no rules about it being unsporting to shoot sitting birds – slaughter was the order of the day, and slaughter it was.'[19]

Members of 617 Squadron also relieved the boredom of no-flying days with games and pranks that showcased their endless – if pointless – ingenuity. One much-prized skill was the ability to put a postage stamp, sticky side up, on top of a two-shilling piece and then flip the coin so that the stamp finished up stuck to the ceiling. Those with a good sense of balance could attempt to do a hand-stand, balance on their head, and in that position drink a pint of beer without spilling a drop. A trick with an even higher tariff required them to stand upright with a full pint on their forehead, slowly recline until they were flat on the floor and then get back to their feet, once more without spilling a drop of beer.

Team games included Mess Rugby, played with a stuffed forage cap for a ball and armchairs as additional opponents to be avoided while sprinting across the Mess. There was also the ascent of 'Mount Everest', which involved piling chairs one on top of each other with 'the odd bod perched in them'. The 'mountaineer' would then climb the heap bare-footed and clutching a plateful of green jelly. When he reached the top of the stack, he placed his bare feet one at a time firmly in the jelly, and then, holding himself upside down, left the imprints of his bare feet on the ceiling. Once the stack of chairs had been removed, newcomers were left to ponder how on earth someone had managed to walk across the ceiling 15 feet above the floor.

Such mountaineering exploits did not always end well. One Canadian pilot on another squadron, Tommy Thompson, covered his bare foot with black ink, but, having successfully made his footprint on the ceiling, lost his balance, fell and broke his leg. He 'played the role of wounded hero quite well' in the Boston and Lincoln pubs, and never told those who bought him a drink why he was on crutches.

The Boston pubs were always packed with aircrew from the surrounding RAF bases, and for those without cars, the scramble to get aboard the last buses that all left from the Market Place at 10.15 every night was, said one airman, 'a sight which had to be seen to be believed'. Most of them just drank beer, and it was 'rather innocuous' because, by government decree, beer was no more than 2 per cent alcohol – less than half pre-war strength – and, says Larry Curtis, 'you had to drink an awful lot of it before you got merry.' There were regular sessions in the Mess, but the aircrew also 'really needed a place away from the station where we could go to forget about the war for a while; we had a lot to be grateful for in the English pub,' Curtis says.[20]

While the officers were enjoying a life of relative luxury at the Petwood Hotel, the NCOs found they had been allocated some rather less salubrious accommodation: a row of Nissen huts and wooden huts erected just outside the airfield's perimeter fence at the side of the B1192 road. Roofed with corrugated iron, they were bone-chillingly cold in winter. One of the NCOs recalled, 'One of the things we had to be careful about, living in Nissen huts in England in the winter, was that if you put your hand against the wall or the roof, it stuck there [because it froze to the metal]. I've seen a few hands with palms left behind.'[21] There was one other peril for crewmen living in the Nissen huts: the trees around them were home to a colony of woodpeckers that drove them mad and disturbed their sleep after night ops with their endless, dawn-to-dusk drilling into the trees with their beaks.

Amidst the camaraderie, the dangers and the tedium of RAF life, the war ground on, and as the Christmas festivities of 1943 faded from their memories, the men of 617 Squadron began to wonder just how long the fight would last.

CHAPTER 4

Death or Glory

A raid by 617 Squadron on a V-1 site in the Pas de Calais on 22 January 1944 – the same day that the Battle of Anzio was launched in Italy – marked another decisive moment in the evolution of the squadron. Leonard Cheshire tried to mark the target from 7,000 feet, but his bomb-aimer, Keith 'Aspro' Astbury – a flamboyant and spectacularly foul-mouthed Australian who was one of the most cherished characters on the squadron – was unsighted by flares bursting ahead of them at the crucial moment and the markers overshot the target.

Mick Martin had previously told Cheshire that he could 'hit a target as small as a clump of seaweed by using his Lancaster as a dive bomber without using the bombsight'. According to another veteran of 617, Cheshire had rubbished the idea at the time, but Martin now set out to prove it. Disobeying his orders, instead of dropping his spot-fire markers from height, he dived down and placed them with precision from 400 feet instead.[1] Subsequent reconnaissance photography showed substantial damage to the site, and the unusually accurate Main Force bombing seemed to justify Martin's claims about the effectiveness of low-level marking.

With the successes they were now achieving, 617 Squadron was no longer being seen as 'the suicide squadron'. They were hitting more targets, losing fewer aircraft, and having considerably more effect than much of Main Force. While the bulk of Bomber Command continued with the policy of laying waste to whole cities, 617 was specialising in the precision bombing of individual targets, a task requiring new techniques and new equipment to produce the spectacularly accurate navigation and weapon aiming that would be required.

The first opportunity for Cheshire himself to test the ultra-low-level target-marking technique in operational conditions came on 8 February 1944, when 617 Squadron was tasked with an attack on the Gnome-Rhône aero engine factory at Limoges in Occupied France, one of a series of 'factory-buster' raids the squadron made targeting crucial links in the German military production chain.

Even with all his experience, Cheshire could still learn from other pilots, and according to one crewman, 'Micky Martin was the "Head Boy" in low level. He taught Mr Cheshire how to fly low level.'[2] To his credit, Cheshire himself acknowledged his debt to Martin, describing him as 'the greatest bomber pilot of the war'.[3] That view was echoed by Martin's crewman, Larry Curtis, who said of him, 'Some idea of the esteem in which I hold him is that I named one of my sons after him – best pilot I ever flew with.'[4]

However, no one, not even Mick Martin, had quite the same degree of coolness and fearlessness over the target as Cheshire. Relying on the Lancaster's gunners to protect their aircraft from fighters, Cheshire backed himself against the flak batteries, flying his own Lancaster in at low level – as low as 50 feet – to mark the target. He also went out of his way to minimise the risk of 'collateral damage' – civilian casualties – on the raids 617 made into Occupied France. However, using the lumbering Lancasters almost as dive-bombers, swooping down to mark a target and executing sharp turns and steep climbs to escape, put huge stresses on their airframes and engines, and it was a tribute to the Lancaster's strength of construction that the aircraft flown by Martin and Cheshire did not fall apart under the strain.

The raid on Limoges was the first op with 617 for Bob Knights' crew, and his bomb-aimer, John Bell, felt:

a frisson of excitement and a real sense of anticipation. This was what I really wanted to be doing: attacking individual installations. It was a crew job to get us all to the right target at the right height and time, but the bomb-aimer had

that final role to make the attack a success and I was certainly conscious of
that sense of responsibility every time we flew an op. On our previous ops
for Bomber Command there were usually hundreds of aircrew at the main
briefing, but now there was only a handful – quite a change!

Regardless of the numbers involved, the ritual preparation for a raid was remarkably similar across Bomber Command. The aircrews all struggled into their flying gear, the gunners wearing long underwear and woollen sweaters beneath their electrically heated flying suits, which were unbearably hot at ground level on a warm summer's day, but vital flying at up to 20,000 feet in their exposed, bitterly cold gun turrets. However, many gunners complained that it was impossible to maintain a steady temperature in the suits. They were protected from the elements only by a flimsy bit of Perspex, but many removed it to improve visibility, and were often sitting in their turrets in a temperature of minus 20 degrees when 'you had icicles hanging down from your oxygen mask'. One gunner recalled it being 'minus twenty-four one night over Berlin'.[5] In order to combat the extreme cold in that position, another gunner plastered all the exposed parts of his face and hands with lanolin, like a Channel swimmer covering himself with grease.

All the crews put on their 'Mae West' life-jackets, with parachute harnesses going over the top of them, and carried their flying helmets or hung them around their necks by the oxygen tubes and intercom cords. They then boarded the crew bus, which lumbered round the perimeter of the airfield, dropping each crew by their regular aircraft. They all looked bulky in their Mae Wests, and the gunners, in their heated flying suits, seemed even larger.

After an external inspection and a word with the ground crew, the pilot signed the Form 700 – the aircraft's Engineering Record Book. One officer recalled the sense of isolation as he waited for take-off on an op one December night, standing around 'warming ourselves at the ground crew's fire, which was burning outside the little shack. It was pretty cold. Things were quiet. No sensation of being surrounded by an air armada waiting to take off. Just a small party in a corner of a big, windy field.'[6]

Before boarding the aircraft, the crews went through their pre-flight rituals, some peeing on the tail wheel, others clutching a battered soft toy or wearing a 'lucky' hat, then they clambered up the ladder and through the narrow hatchway by the rear wheel, dragging their flight bags and parachute packs with them. For such a large aircraft, the interior was remarkably cramped – six men jammed into a space no bigger than the interior of a small van, with the seventh in lonely isolation in his turret at the tail – but then, the design priority had been room to house the bomb-load, not the crew. Even relatively short men had to stoop, and in their bulky kit it was a struggle to move along the narrow passageway and over the main spar of the aircraft; for a tall man like bomb-aimer John Bell it was a constant trial by ordeal, banging his head, scraping his knees and catching his clothing on protruding metal.

Take-off for the attack on the Gnome-Rhône engine factory was at nine that cold February night, and the twelve crews held off, circling at a distance from the target, while Cheshire and Mick Martin carried out the target marking. The factory employed 2,000 French civilians, mostly women, and because of the risk to them and the damaging propaganda that would ensue if large numbers were killed in a raid, it had actually been struck off Bomber Command's list of potential targets. Two constraints had therefore been imposed on 617 Squadron: no French civilians were to be killed and, to ensure that and to maximise the damage to the factory, all bombs had to fall within the target area. One of the Lancasters was adapted so that a cine-film could be made of the raid, in the hope of providing evidence to contradict any German propaganda claims, and to generate British propaganda if the raid proved successful.

As a result of those constraints, Cheshire first made three low-level runs across the factory to give the workers inside – 'mostly French girls'[7] – warning that bombs would soon be falling on it. It was a difficult, twisting approach down a narrow valley, swerving around two tall water towers and a factory chimney, but Cheshire and Martin then placed their yellow incendiaries squarely in the middle of the factory, their accuracy helped by the fact that the lights in the factory were blazing, with the black-out both there and in the town itself 'very poor'. As the factory lights were extinguished, Cheshire called in the remainder of the squadron to bomb the burning markers. 'I couldn't really see the target itself most of the time,' Johnny Johnson said. 'You bombed the different coloured flares that Cheshire had dropped. So he would go in, drop the flare on the target, then radio to instruct us to bomb a certain coloured flare.'[8] The fires of the flickering markers lit up the surrounding factory buildings and cast an eerie glow across the site.

The new low-level marking technique proved devastatingly effective. All but one bomb landed inside the factory compound, and so tight was the bombing that the blast from one 12,000-pounder almost extinguished the fires started by Cheshire's incendiaries. The factory was virtually wiped off the map. 'We flattened the target but saved the civilians,' Johnson says. 'That gave us a real sense of accomplishment. Cheshire held us back until he thought they were all clear, and later he got a letter thanking him for ensuring their safety!'[9]

In fact Johnson may be remembering a subsequent raid on a factory in Angoulême on 20 March 1944, after which they received a message from the mayor, thanking them for not killing any French people, though the mayor added that he couldn't understand 'why the British were firing at the street when the French were coming out and cheering us on'. But Mac Hamilton recalled, 'As always in 617, once you dropped your bomb, you just didn't beetle off home, you went round again and shot up flak towers and distracted the gunners while the other aircraft were coming in to bomb.'[10]

The morning after the Limoges raid, the 617 aircrews were told the attack had been a success, and most of them realised at once that it was a defining moment. A still from the cine-film of the raid, showing Cheshire's incendiaries landing on the factory roof, was released to the world's press, adding fresh lustre to the grow-ing legend of the Dambusters. 'We sensed that this type of target marking heralded a new phase of the war for us,' John Bell says. 'It was good to be away from the mass bombing.' He had had plenty of experience of area bombing.

★   ★   ★

Even today, Bell remains tall and sharp-featured, with a high fore-head and a keen, penetrating gaze, and, now entering his nineties, he still holds himself ramrod-straight. He was only sixteen years old when war was declared on 3 September 1939. During the summer he had heard his father and friends talking about the pros-pect of war but, Bell says, 'it all seemed unreal. We thought that diplomats would talk our way out of the war. It couldn't really happen, could it?'

His father had served in Egypt during the Great War but never spoke about it and, like most people then, didn't expect the war to last long. He thought that all the men rushing to join up would soon be back looking for jobs, so he told his son he'd be better off starting a career before the competition got too intense. Bell began training as an accountant but, not wanting to be left out of the war effort, he also joined the Home Guard, 'so I was doing a bit of parading around with a rifle and no bullets,' he says with a smile. He was also trained in sabotage in case the feared German invasion took place.

Bell was living in Surrey and working in London, and once the 'phoney war' had ended with the onset of the Blitz, he had a close-up view of the impact of the war. One night as he was walking home, some bombs fell close by and he ended up shel-tering under a park bench. 'The air-raid sirens were howling, the anti-aircraft guns were firing and shrapnel from the shells was falling all around me,' he says. 'I think that was more dan-gerous than the German bombs – chunks of hot metal fizzing down around you. It was certainly scary, but there was nothing you could do about it – it was happening everywhere.'

However, seeing its effects so close at hand made him deter-mined to play his full part in the war. 'I'd seen the Army come back from Dunkirk in tatters,' he says, 'watched the Battle of

Britain raging over my head, and seen the results of the bombing: the destruction, flattened buildings and smoking ruins. It made me want to be part of the fight, to hit back at the Germans who were attacking our country.'

Like all recruits to the RAF, Bell had hopes of being a pilot but they were soon dashed. At six feet four inches tall, he was 'deemed to be too long in the leg' to get out of a cockpit in a hurry and began training as an observer and navigator in South Africa. He knew nothing of the country other than what he had read in magazines, and 'imagined jungles and wild animals everywhere', he says smiling. 'We were all wide-eyed boys going out there, very naive!'

He returned to England as a fully trained observer, but was then informed that he would be a bomb-aimer instead. Being so tall always made it awkward to get into his bomb-aimer's position, but 'twenty years old and skinny, I could manage to get through a lot of difficult places'. Even when he'd got himself into position, it remained an uncomfortable experience for Bell, with no room to stand upright in the compartment. Fortunately, the bomb-aimer was the only member of the crew who could see directly beneath the aircraft through the bulbous Perspex 'goldfish bowl' in the nose, so he had a grandstand view of the bombs going down and the flak coming up.

After completing his training, he went to an Operational Training Unit where he 'crewed up'. Pilots, navigators, wireless operators, bomb-aimers and gunners were all assembled in a hangar and then told to form into crews. Bell was talking to a Canadian navigator, Harry Rhude, when a much older man, a rear gunner, joined them and said he'd found a pilot. 'I know he's had a crash during his training,' he said, 'so I think he'll be a bloody sight more careful in future!'

The rear gunner introduced them to their pilot, Bob Knights, with the words 'I've found you a navigator and a bomb-aimer.'

'Oh good,' Knights said. 'All we need now is a wireless operator,' and promptly went off to look for one. 'It seemed pretty haphazard,' Bell says, 'but I don't think there was any other way to do it. My only thought was: You are choosing the people you'll spend the next few years with, live with, possibly die with. So who would you trust most?'

Having crewed-up, they were sent to a Heavy Conversion Unit where they flew a Lancaster for the first time. Bell liked everything about the aircraft except the long trek back to the Elsan chemical toilet at the rear of the aircraft. 'It was a good aircraft, very robust, and never really gave us any trouble.' The Elsan was obviously not an option for the pilots, who had to make alternative arrangements. The Australian pilot Bruce Buckham's crew 'very kindly kept the tops of the smoke floats we were tossing out', so if his bladder was bursting, he'd use one of those, then open the chute they used for dropping Window, and the suction was so strong that it would go straight down the chute. 'Unfortunately the second time I used it,' Buckham recalled, 'I spilled some and it went all over the bomb-aimer down below. He was not pleased. That's where the expression comes from – though Guy Gibson introduced it – "pissed on from a great height"!'

John Bell's crew were posted in June 1943, joining the newly formed 619 Squadron. As they arrived at their base at Woodhall Spa, Knights murmured, 'I wonder how long we're going to last here.'

'I remember that to this day,' Bell says.

It was an off-the-cuff remark, but we knew the losses that were being suffered by Bomber Command, though I was quite surprised to hear Bob referring to it. As a rule we never discussed losses, and every time we heard of them, it was always a number of aircraft, not people. I'd hear on the radio 'Bomber Command lost thirty aircraft last night,' but I never translated that into numbers of people at the time. Years later I did think about it, knowing that thirty aircraft meant over two hundred people had been killed. You'd see empty spaces at breakfast or beds being cleared away, but you didn't let it affect you. We all had the same attitude: It won't happen to me. Of course, later on, I realised that there were around fifty-five thousand men who had said the same thing, and it did happen to them.

Their aircraft's designation letter was T, and Knights, who had recently seen the Disney film *Bambi*, released in 1942, christened his aircraft 'Thumper' and had Bambi's rabbit friend painted on the nose. Their first op in T-Thumper on 24 July 1943 would have been memorable to them for that reason alone, but it also happened to be the launch of Operation Gomorrah – a series of mass bomber raids on Hamburg by Bomber Command and the USAAF that in the course of eight days and seven nights effectively destroyed Germany's second city. It was also the first time that Window was dropped to give false signals to German radar, which was 'a bit of luck for me', Knights later said, 'because the bombing was more or less unopposed.'

John Bell's crew before a raid on Frankfurt

They'd flown out over the featureless darkness of the North Sea and the blacked-out landscape of Germany. The first lights John Bell saw in the far distance were the beams of searchlights piercing the night sky over Hamburg, the flashes of exploding flak shells and the glow of the drifting smoke from the shell-bursts as it was caught and illuminated in the searchlight beams. As they flew closer, he saw the first fires and burning buildings on the ground from the bombing force ahead of them, but also the mass of flak-bursts through which they would also have to pass:

Stuck in my Perspex bubble in the nose, surrounded by nothing other than flimsy plastic that offered no protection at all, I had the best view of the flak. It looked pretty threatening, and someone would always say: 'Looks like the natives are a bit unfriendly tonight!' The flak barrage was in full swing when we arrived and we had to fly into that – me first! It was lighting up the residual smoke, so it looked both alarming and spectacular – it may have been dangerous but you just have to get on with the job. I was apprehensive, but I don't remember any real fear. I just thought, How are we going to get through that? Then I just concentrated on the bombing run and ignored everything around me.

He had a clear view of the bombing's impact on the city below them. 'I'd seen the impact of bombing close up in London, but looking down on this mass of burning buildings was my first sight of what Bomber Command could do, and it was an awesome – and awful – sight. I didn't think about the people on the ground at that time – I did later – but back then, it was all just part of the war. When we came out the other side, I just heaved a sigh of relief and told the pilot to climb higher and get out.'

At the end of the week-long Operation Gomorrah, Hamburg had been almost completely destroyed. The hot, dry weather and the mainly wooden construction of the houses fuelled the firestorm ignited by thousands of tons of incendiaries and high-explosive bombs. Generating temperatures of 800° C and wind-speeds of 150 miles an hour, it created a ferocious vortex of fire that rose over 1,000 feet into the air and swept across the city, consuming everything in its path. The tarmac of the streets burst into flame, and spilled fuel oil ignited as it spread across the surface of the canals and harbour, making it seem as if even the water was on fire. Even air-raid shelters and deep cellars offered no protection; people sheltering in them were suffocated as the firestorm consumed the oxygen. Operation Gomorrah killed over 40,000 people and left a million more homeless.

Of all the Main Force ops that Bell flew, he retains the strongest memories of that first raid on Hamburg, though he also vividly recalls Berlin, because it was so heavily defended and they attacked it so often. The impact of one of those mass raids on Berlin in November 1943 was vividly described by a Swedish businessman who found himself trapped in the city as the bombs fell. His account gives a powerful insight into the horrific experiences of German civilians pinned under Bomber Command's relentless onslaught:[11]

'The fire brigades and ARP personnel are powerless to cope with the situation. Day has been turned to night by the billowing clouds of evil-smelling smoke which fill the streets. The sky is blotted out.' The Ministries of Propaganda and Munitions were badly damaged, the Foreign Office in the Wilhelmstrasse was wrecked, as was the gigantic Air Ministry building in Leipzigerstrasse – 'Göring's pride and joy'. The Wilhelmstrasse and Unter den Linden districts were burning so ferociously that 'firemen have given up the hopeless struggle. They have cordoned off whole blocks of buildings and simply left them to burn themselves out. Armed guards equipped with gas masks against the suffocating smoke are stationed at the cordons.'

The once beautiful, tree-lined Unter den Linden was 'a shambles', with almost every building on fire. 'There was a sound of hissing as light rain fell on the flames.' The University State Library and the Bristol Hotel – one of Berlin's finest – were destroyed. The Adlon Hotel, requisitioned for the homeless, was still standing, but all its windows had been blown out. The Gestapo headquarters in Prinz Albrechtstrasse and the headquarters of the Berlin police were both badly damaged. There was an SS cordon round the workers' quarters north of the Alexanderplatz to prevent workers leaving the factories and escaping to the country, and armed guards also surrounded Berlin's zoo in the Tiergarten, while troops armed with rifles and machine guns hunted the leopards, elephants, bears, tigers and lions which had escaped after the zoo was hit by bombs. 'Berliners, fatalistic, now believe that the RAF will return every night until Berlin is in ruins.'

★   ★   ★

Night after night, 617 Squadron returned to occupied Europe, and on the evening of 12 February 1944 twelve Lancasters took off to attack the Anthéor viaduct on the French Riviera. Five hundred and forty feet long, it spanned the gap between two rocky headlands a few miles west of Cannes on the Côte d'Azur. They had already bombed it twice before, in September and November 1943, but on both occasions it had sustained only slight damage to the railway tracks and virtually none to the viaduct itself. The bombs they dropped were delay-fused because, flying so low, if the bombs exploded on impact, the blast would destroy the Lancasters as well as the viaduct, but the ground around the viaduct was solid rock, and as a result the bombs were 'bouncing all over the place' and the chance of any of them coming to rest close enough to one of the columns of the viaduct to destroy it was very remote indeed.[12] Nonetheless, they were to make one more attempt, because the line the viaduct carried, a vital rail link between the south of France and Italy, was now being used to transport 15,000 tons of military supplies a day to German forces resisting the Allied advances from the Anzio bridgehead in Italy.

At the pre-flight briefing, the intelligence officer had warned them that the viaduct was defended by twelve heavy guns, several lighter ones and searchlights, though there would at least be no moonlight – on one of the previous raids, the moonlight reflecting from the surface of the Mediterranean had made it seem 'almost as light as day', making the raiders clearly visible to the flak batteries. Carrying 12,000-pound Blockbusters in their bomb-bays, the target was at the extreme limit of the Lancasters' range.

They crossed France at low level, climbing to height to cross the Alps. It was a bitterly cold night, with 'quite severe ice' forming on the aircraft. As on the previous ops against the viaduct, cloud was obscuring the area, making it difficult to mark the target with spot fires from high level. The ground defences were also much more active and organised, with banks of searchlights and 'pretty damn heavy' flak.[13]

Cheshire and Martin were to illuminate the target to allow the rest of the force to drop their bombs but, dazzled by searchlights,

Cheshire could only drop his markers on the beach at the entrance to the inlet where the viaduct was sited. Martin dived to low level, swooping down the valley towards the viaduct. Even flying exceptionally low, he also had problems from 'an awful lot of searchlights which made it very difficult to see anything'. The bomb-aimer, Bob Hay, squinting through his bombsight, had just said, 'I've got her, Mick,' when the Lancaster was hit 'very hard by anti-aircraft fire', knocking out two engines.[14] A flak burst ripped through the starboard side of Hay's compartment and there was no further response from him. The flight engineer, Ivan Whittaker, was also badly wounded and lay doubled up on the floor of the aircraft, having been hit in the leg.

Struggling to control the aircraft, Martin aborted the bomb-run. As he turned, Nick Knilans in R-Roger dropped down to 3,000 feet and his gunners opened up on the searchlights – 'they all went off'. He climbed back to 10,000 feet and made his bomb-run towards the viaduct, which was silhouetted against the surf-line along the beach. In turn, he and the other raiders dropped their bombs and turned for home, but the difficult approach to the target, the bristling air defences and the rocky terrain that made their bombs bounce away from the viaduct defeated them yet again, and it remained intact.

Mick Martin's aircraft was too badly damaged for him to return to base. The electrical systems were damaged and the bomb-release mechanism was jammed, keeping the bombs in the bomb-bay with the doors hanging open. With only two engines, Martin struggled to prevent the heavily laden and badly damaged aircraft from hitting the sea. Above him, looking down in the bright moonlight, Les Munro could see 'Mick down so low over the water, he was creating a wash on the surface from the force of his props'.[15] The propellers were whipping up such a mist of spray that Martin's crew at first thought it was smoke from an onboard fire.

Although the bomb-release circuits were destroyed, the crew eventually managed to jettison their bomb after the navigator prodded the release mechanism with his long navigation ruler. The bomb detonated as it struck the water and 'there was quite some wham', lifting the aircraft 'like an escalator after the start button was pushed'.

With no hope of making it back to England, Martin set a course across the Mediterranean while one of his crew tried to work on Bob Hay's wounds, but in the pitch darkness inside the aircraft it was almost impossible to see what he was doing. Wireless operator Larry Curtis sent an SOS to a fighter airfield on Corsica, but when he asked for a doctor to be put on standby, he was told that none was available. Martin then made for Sardinia and Elmas Field, near Cagliari. He made an emergency landing there, stopping only inches from the end of the small runway, beyond which was a sheer drop into the sea. By then his bomb-aimer, Bob Hay, a veteran of the Dams raid, was already dead. 'A single shot came through the nose and killed him; that was his goodbye.'[16] He was buried in a rough pine coffin in the local cemetery, a few feet from two German graves.

Mick Martin's had been one of the first all-Australian bomber crews, and they had been flying ops together since October 1941. Martin had always said that they would 'fly as a crew, live together as a crew, and rely on each other from head to toe, high or low, on ops or on leave'. Such was the loyalty to each other of that original crew that, though they had continued to fly ops after completing their required tours of duty, they pledged that if any of them were killed, the remainder would 'take the hint' and come off operations together. True to their word, the Anthéor raid in which Bob Hay died was the last op that Mick Martin, Tom 'Tammy' Simpson and Toby Foxlee ever flew with 617 Squadron. As soon as he returned to Woodhall Spa, Martin told the adjutant Harry Humphries that 'he would now have to leave the squadron. It seemed an unbelievable decision, but he meant it.'[17] With Martin standing down, Larry Curtis joined Joe McCarthy's crew and was also made the squadron's signals leader. That Martin had not stood down because he'd reached his physical or mental limits was demonstrated when he immediately joined 515 Squadron and went on to fly another thirty-four operations, piloting Mosquito fighter-bombers.

The other crews had all landed safely at Ford in Sussex to refuel, but on the short final hop from there to Woodhall Spa there was another tragic loss, as the Lancaster of a Canadian pilot, Squadron Leader Bill Suggitt, crashed into a hill on the Sussex Downs soon after take-off with the loss of all eight men on board. Among them was Squadron Leader Tommy Lloyd, the squadron's intelligence officer and a veteran who had flown with the Royal Flying Corps during the First World War. Suggitt, a non-smoking, non-drinking bookworm with a stand-offish, almost condescending manner, was not universally popular on the squadron, but he was a gifted pilot who could 'throw a Lancaster around as if it was a child's toy'. Lloyd was the polar opposite: good-humoured, gregarious, he 'threw parties, stood all the expense, took his drink like a gentleman and yet still found time to work more hours than necessary'.[18] Now both were gone.

Although, as usual, the losses were not discussed, they were nonetheless keenly felt. 'I think it had more of an effect on us,' John Bell says, 'because we were a closer-knit squadron. On Main Force, a crew was just a pilot's name on a board. Here, you knew crews better and mingled with them more.'

A week after the Anthéor raid, the USAAF and Bomber Command's Main Force jointly launched Operation Argument – less formally known as 'Big Week' – which aimed to lure the Luftwaffe into a decisive battle by launching mass attacks on Germany's aircraft-manufacturing plants. It was an American plan and Bomber Harris had initially refused to allow his aircraft to take part, since he felt it was diverting resources from the continuing area-bombing campaign against German cities, but pressure from the Chief of the Air Staff forced him to comply. Although it failed to strike a decisive blow and US losses of 250 bombers and 30 fighters suggested that Harris's reservations might have been justified, German losses, particularly of fighter pilots, were a major blow to Germany's air defences, and from then on, Allied air superiority grew ever more pronounced as the war continued.

★   ★   ★

Meanwhile, ten new crews had arrived to bolster 617 Squadron's depleted ranks, with each of 5 Group's other squadrons contributing one top crew each. Among them were crews captained by John Pryor from 207 Squadron, Mac Hamilton from 619 Squadron and Don Cheney from 630 Squadron.

Don Cheney in training

As far back as he could remember, Don Cheney had always wanted to fly. He still has the scrapbooks he filled as a child with pictures of aeroplanes that he'd drawn, and underneath each one are notes of its maximum speed, height ceiling, armaments and so on. 'I built an air force on paper,' he says. 'I used to dream about flying and I even used to dream that I could fly!'[19]

He grew up in Canada and was seventeen when war broke out. Two days before his eighteenth birthday he signed enlistment papers and fulfilled his childhood dreams by joining the RCAF as a trainee pilot. 'I was an only child and my parents were desperately worried about me leaving for war. They did their best to keep stiff upper lips, but my dad's heart must have ached, because he knew that we were in for another conflict as bad as the "war to end all wars" that he had fought.'

On 9 October 1942, Cheney said goodbye to his parents and his girlfriend Gladys at Ottawa Union railroad station. It was a sad parting but, says Gladys, 'It was the reality of war and the same for everyone – so many people were saying goodbye to one another back then. You just hoped you'd be reunited one day.' For the next two years she had no idea what he was doing, or even where he was. 'We didn't get that sort of information,' she says. 'The letters he sent were subject to censorship and could give only the most general and bland information about what he was doing. So I just had to hope and pray that he was safe.'

He arrived in Britain with 12,000 other Canadians on board the *Queen Elizabeth* and was sent to Bournemouth for further training, joining hordes of Commonwealth aircrews waiting impatiently to join the fight. Almost every day a German Focke-Wulf Fw 190 would come roaring in at low level from the Channel, strafe the buildings and drop 500-pound bombs. 'One feller close to the explosions was actually blown off the john!' recalls Cheney with a laugh. 'This was my first sight of what the war would mean and it put the wind up all us new guys; I realised that they really meant it – this wasn't Canada any more!'

Cheney began flying Lancasters with 106 Squadron. He was still so boyish-looking it was a wonder he was allowed in an aircraft at all, let alone fly one, but he proved a very capable pilot. To complete the crew, they recruited a flight engineer, Jimmy, a Scottish mechanic. On their first flight, Cheney took off and called, 'Wheels up!' There was no reply. 'Wheels up, Flight Engineer! Wheels up!' There was still no response.

When Cheney glanced across, he saw his flight engineer with a greenish tinge to his face, gripping the bar across the window as if his life depended on it and staring in horror at the receding ground below him. It turned out that Jimmy had never flown in an aircraft before. Cheney jabbed him in the shoulder and said, 'Get those bloody wheels up!' The flight engineer tore his terrified gaze away from the view through the window and did as he was told. A few moments later, one of the starboard engines caught fire. They circled once and made an emergency landing with flames trailing from it. 'Out came the Fire Brigade, doused us with foam, and that was our first day over – never a dull moment!'

When they weren't flying or the weather was bad, Cheney and his crewmates, like most other aircrew, headed straight for the pub. 'We could all sit down, enjoy company, take our minds off the war and meet some young ladies,' he says. He and his wireless operator clubbed together to buy a maroon Austin 10 convertible they saw advertised in a newspaper. They caught a train to Croydon, paid seven pounds for the car and set off back to base through the centre of London.

The car didn't prove to be such a bargain and broke down right in the middle of Trafalgar Square. Resplendent in full service uniform, they were trying to change the distributor head while traffic piled up behind them, when a large British bobby appeared, tapping his truncheon in the palm of his hand, and said, 'Move it! Now!' Sweating and cursing, they had to push the car out of Trafalgar Square. Once repaired, the Austin became the crew transport to pubs, dances and anywhere else they wanted to go. 'We could get all seven crew in there,' Cheney says, 'and more if we needed: three in the front, three on the back seat, and three sitting up on the folded-down hood. Once we were heading for the pub and I took a corner a bit too fast. The three fellows sitting on the folded hood were thrown clear off and ended up in the ditch at the side of the road!'

Cheney flew eighteen ops with 106 Squadron and saw plenty of aircraft shot down. 'Of course, at night you couldn't see the aircraft themselves,' he says:

You'd see the flashes of flak bursts, a ball of sparks with a bright centre, then falling flames and an explosion. It was a stark sight – I realised this was fellow aircrew dying – but there was little time to dwell on it. Getting to the target was the most important thing to do, the sole reason we were all in the sky, so you just pressed on and prayed that the Lord was on our side, riding on our tail.

On moonlit nights, he says, it felt like 'a shooting gallery for night-fighters'. There was also a risk of being 'coned' by searchlights. If one or two searchlights caught a Lancaster in their beams, others would swing on to it as well, locking it in a cone of light that was so bright that the pilot often literally could not see anything at all. 'You were totally dazzled, and once you were coned, the flak opened up on you and frequently that was that.'

> *Fortunately we were only coned once with a full bomb-load aboard. I stuck the nose straight down, dived down for five or six thousand feet, then twisted, turned and heaved with all my force on the control column in a climb back to near stalling speed and then pushed the nose back down again. Each time I was weightless for several seconds, my bum lifted right off the seat and I was left hanging in mid-air in my safety straps. It was all I could do to level out and push the controls forward again but, unlike a lot of other guys, we managed to get out of it.*

The same week that Cheney had joined 106 Squadron, its commander, Guy Gibson, had left to form 617 Squadron for the Dams raid. Now Cheney's crew were about to follow in Gibson's footsteps, though when he received orders to report to his Station Commander in the summer of 1943, he at first assumed he must have done something bad and was about to be sent home in disgrace. That fear disappeared when the Station Commander greeted him warmly, told him to sit down and relax, gave him a cigarette and even lit it for him. He informed Cheney that 617 Squadron was rebuilding after the losses on the Dams raid and wanted his crew to join them.

Cheney's first thought was 'My God, 617 Squadron – "The Death or Glory Boys"' – one of the many nicknames they had been given by the rest of the RAF. He asked for time to discuss it with this crew and they met in his room over a few beers. A couple of them were 'somewhat reticent about 617', Cheney says.

> *We knew about their losses and were fully aware of the dangers involved, especially in relation to low flying, but the prestige of the squadron won the day, especially because of the targets they were attacking. This wouldn't be about dropping tons of bombs on cities or bombing somebody's house, we would be targeting specific aspects of the enemy war machine: factories, weapons sites, bridges. It was our chance to make a real difference.*

When they eventually arrived on 617 Squadron, they were given a vivid demonstration of the dangers they would face, for it was only a few days after the raid on the Anthéor viaduct in which Mick Martin's bomb-aimer, Bob Hay, was killed, and Bill Suggitt's Lancaster crashed with the loss of all of the crew. It was a sobering welcome to their new squadron, and the coming weeks would provide plenty more reminders of the dangers they all faced when, during an operation against the aero-engine factory at Woippy on 15 March 1944, the actions of one air gunner contributed to the growing legend of 617 Squadron.

★   ★   ★

Flight Sergeant Tom McLean had already completed a tour flying Halifax bombers with 102 Squadron and was on his very first op with 617 Squadron.[20] Just under six feet tall, with a shock of curly hair, he had a lean, rangy build and a broken nose as a souvenir from one of a number of occasions when he'd settled an argument with his fists, not his words. A dour Scot from Paisley and a boxer who had represented the RAF, he was not always the easiest of companions, but there was probably nobody in Bomber Command who had thought so deeply about German fighter tactics and the operational gunnery techniques to counter them, nor shown such expertise in putting theory into practice.

He studied, and relentlessly practised, every aspect of air-to-air gunnery, including range estimation, accurate sighting and

deflection shooting (aiming ahead of a moving target so that
the bullet and target arrive at the same place at the same time),
and he worked closely with the pilot and the other gunners in
his crew on evasion tactics, which they then practised assidu-
ously on training flights. His attention to detail covered every
aspect of air gunnery and he even assembled his own ammunition belts, believing that the standard Bomber Command 'mix'
of successive ball, tracer, incendiary and armour-piercing rounds
was less effective than a mix of 40 per cent tracer and 60 per cent
armour-piercing.

McLean proved the effectiveness of his methods during a night
raid on Mannheim on 6 December 1942. He was the rear gunner
of a Halifax when a Junkers Ju 88 night-fighter opened fire,
wounding McLean in the left hand. He and the mid-upper
gunner immediately returned fire, riddling the port engine and
wing of the fighter and sending it spiralling down in flames. Two
more fighters then launched a coordinated attack, but both gunners again returned fire, destroying another of the enemy aircraft
and forcing the last one to break off the attack. The Halifax
carried on to its target, and on its return to base both McLean
and his fellow gunner were awarded immediate DFMs.

During his tour with 102 Squadron, McLean had confirmed
'kills' on five enemy fighters, with a sixth unconfirmed one,
and his exploits made him the subject of a profile in the *Daily
Mirror* under the cringe-making headline 'They Call This Gunner
"Killer".'[21] No one ever had before, but his crewmates took great
delight in calling him 'Killer' after that!

His successes in air combat had vindicated his theories, but his
attempts to convince senior officers of their merits proved less
successful. When he suggested to his squadron's gunnery officer
that other gunners should switch to his mix of tracer and armour-
piercing rounds, he was told that the make-up of ammunition
belts had been decided by the Central Gunnery School and could
not be altered on the word of a mere air gunner – and an NCO,
to boot – and in any case too high a mix of tracer would damage
the rifling of the gun barrels. When McLean countered that gun

barrels were easier to replace than dead aircrew and shot-down bombers, he was dismissed with a waft of the officer's hand. Astonishingly for the era, he simply went over the senior officer's head by writing a paper outlining his experimental methods and the results he obtained and sending it to the Group Gunnery Leader, but the only reply was a stern reprimand, reminding him of the absolute necessity of only submitting correspondence through 'the proper channels'.

After ending his thirty-op tour with 102 Squadron, he was posted to RAF Alness in Invergordon as a gunnery instructor with Coastal Command. His generally 'awkward' reputation may have contributed to the decision to send him to what must have seemed to him the British equivalent of Siberia. Other men might have been delighted at such a cushy posting, far from the heat of Bomber Command's battles over Germany, but McLean, who had arrived at Alness eager to impart his knowledge to other air gunners, became so frustrated and so furious at the indifference of his commander and the other permanent staff that he was even contemplating deserting when the chance to join 617 Squadron presented itself. A former comrade, who had served under Leonard Cheshire, urged McLean to write to him, listing his expertise and achievements and applying to join the squadron.

It was unusual for men to apply to 617 Squadron, but Cheshire at once saw in McLean's letter the expert knowledge and passionate commitment that could add an even keener cutting edge to his squadron, and within five days of writing it, McLean was on his way to Woodhall Spa. To such a man, the maverick atmosphere of 617 Squadron, where room could be made for any 'difficult' character as long as he had sufficient expertise in the business of flying and fighting, was as welcome as a warm fire on a cold winter's night.

Cheshire paired McLean with a crew of 'wild colonial boys' led by a Canadian pilot, Flying Officer Bill Duffy. McLean's ideas on evasion and counter-attack tactics against night-fighters were warmly welcomed by Duffy, and they practised them in a series of training flights. The problem for all air gunners facing German night-fighters was that the effective range of their Browning .303 machine guns was only 300 yards, whereas the fighter's 20mm cannon could be lethal from twice that range. As a result, German fighter pilots could lay off out of the Browning's range and direct cannon fire into the bomber they were targeting in perfect safety. When targeted by a predatory fighter, McLean's solution was for his pilot to lure the fighter into the Browning's effective range by combining the traditional 'corkscrew' evasive manoeuvre with a violent deceleration. The pursuing fighter pilot would normally expect a bomber to accelerate in an attempt to escape, and by the time he realised that he was suddenly closing on his prey at twice the expected rate, the Lancaster's gunners would be opening up on him. As soon as Duffy heard the thunder of his own aircraft's guns, he was free to 'pile the coal on again', pushing the throttles wide open and accelerating away.

They practised the manoeuvre repeatedly on training flights over the Wash, with McLean constantly urging Duffy into ever more violent corkscrews and savage decelerations. Their first chance to test how effective their system would be under combat conditions came in the raid on the Woippy aero-engine factory, just north-west of Metz, on the night of 15 March 1944.

On their outward flight, they crossed the Channel coast at Beachy Head, and soon afterwards McLean and the mid-upper gunner, 'Red' Evans, test-fired their guns into the grey waters below them. McLean had personally loaded his ammunition belts that afternoon, filling them with his patent tracer/armour-piercing mix, and he and Evans kept a constant watch on their sectors of the sky, Evans checking the port and starboard beams while McLean, in his rear-facing seat, scanned the sky behind them, the most likely source of attack.

They were only half an hour from Woippy when they received a 'boomerang' recall order because of dense cloud cover over the target, giving them no chance of accurate bombing. At once, Duffy banked the Lancaster around and headed for the safety of home.

Soon afterwards the wireless operator reported that he was picking up a signal from the aircraft's Monica set – radar warning equipment that alerted the crew to enemy aircraft approaching from the rear – indicating an aircraft about 1,200 yards off the port quarter. (Monica was withdrawn in summer 1944 when the RAF discovered that the Germans had a radar receiver that could home in on it.) McLean raked the area with his gaze until he spotted a dim shape faintly outlined against the night sky. It appeared to be a four-engined aircraft, and Duffy and most of the other members of the crew relaxed, believing it was another Lancaster, but there was something about its profile that rang alarm bells in McLean's head and he kept his gaze fixed on it.

He had a further cause for alarm when Duffy gave Red Evans permission to leave his post in the mid-upper turret to use the Elsan toilet at the rear of the aircraft. If the following aircraft turned out to be a fighter, McLean would have to fight it off single-handed until Evans returned to his post, and since Evans had to disconnect his intercom on leaving the turret, there would be no way to recall him until the thunder of guns gave the alarm.

McLean kept staring at the dark shape. It was steadily closing on them, and within a few more seconds he was sure. He flicked his microphone switch. 'Skip, that's not a Lanc following us, it's two enemy fighters in formation! They're out on the port quarter at about nine hundred yards! Prepare to corkscrew port!'[22]

The wireless operator then stood up in the astrodome and spotted a third fighter, a Messerschmitt 109, flying about 500 yards off the port beam with its white navigation lights flashing. It seemed to McLean that it was serving both as a distraction to the bomber crew and as a visual reference point for the other fighters, in case they lost the bomber while manoeuvring in the darkness. A moment later one of the two fighters behind them launched its attack. McLean waited a couple of seconds and then shouted, 'Go!' Duffy at once threw the Lancaster into a violent corkscrew to port, while McLean hunched over his guns, taking up the first pressure on the triggers as he waited for the Messerschmitt to come within range.

The fighter, an Me 110, opened up with its machine guns and cannon from about 650 yards away. The cannon fire passed over McLean's turret but the machine-gun bullets shattered some of the Perspex of his turret and chewed their way through the rear fuselage.

Fortunately for Evans, he had leapt off the Elsan and begun to battle his way back to his turret as soon as Duffy had thrown the aircraft into its corkscrew. Had he remained where he was a few seconds longer, he would have been cut to pieces by the machine-gun fire.

The fighter closed to within 500 yards, then made a diving turn to port and took up station alongside its companion again, but a few moments later it returned to the attack. McLean called for another corkscrew to port and the violent deceleration they had practised so often. This time, perhaps encouraged by the lack of gunfire from the Lancaster, the fighter closed to within 300 yards. Even though Evans had not yet returned to his turret, it was the chance McLean had been waiting for. As the Lancaster's violent evasion again saw the fighter's cannon fire passing harmlessly over the top of the fuselage, McLean opened fire, the clatter of the spent casings drowned in the thunder from the Browning's quadruple barrels. He saw the streams of tracer burning their way across the fighter's port wing and engine as the armour-piercing rounds tore it apart. Within seconds flames and black smoke were belching from the engine. The damaged fighter was still closing, however, and McLean called 'Drop!' into his mic and then put a few more short bursts into the fighter's cockpit. At once the Me 110 reared up and then plunged down in a death-spiral, still trailing smoke and flames.

The danger was far from over, and McLean at once switched his gaze, locking on to the second enemy fighter, still lurking on the port quarter. Evans had now struggled back into his turret, announcing his presence with a startled exclamation of 'Jesus!', followed by a burst of tracer at the Me 109, which was still showing its navigation lights. In a few terse words over the radio, McLean explained the situation and drew Evans's attention to the Me 110 to the rear, telling him to hold fire until he himself had opened up.

The Me 110 launched its attack by flipping onto its back, then curving in towards the Lancaster as Duffy threw it into yet another vicious corkscrew. The fighter kept coming, closing to a suicidal 250 yards. Both aircraft opened fire simultaneously but the Messerschmitt's cannon fire again passed above McLean's turret, whereas his Browning stitched an unerring line right along the underbelly of the fighter. Red Evans's guns now also found their mark. The fighter's momentum carried it above the Lancaster, but McLean swung his gun barrels to cover it and put another short burst directly into the cockpit. The pilot's head was outlined for a moment by the fierce glow of the tracer and then his aircraft broke up in mid-air. A series of internal explosions blew it apart as it fell to earth.

The Me 109 was still holding station, suggesting the danger was not yet past, and soon afterwards McLean spotted yet another Me 110 ranging up on the port quarter and then dropping in almost directly astern at a range of 700 yards. As Duffy saw a bank of cloud ahead and below them, and accelerated towards it, he triggered the fighter's attack. On McLean's shouted order of 'Go!' Duffy abandoned the attempt at concealment and launched a corkscrew to starboard, then on the command 'Drop!' he jerked back the throttles. The Lancaster must have appeared to stop dead in the sky to the fighter pilot, and as he overshot, his own cannon missed the mark, leaving him exposed to McLean's pitiless fire. The Brownings spat a line of tracer and armour-piercing rounds that raked the fighter from nose to tail. Its port fuel tanks erupted in a fireball, and flames also gouted from the starboard engine. McLean ceased fire, knowing his adversary was already finished. He had momentary glimpses of sparks arcing from the severed cables of the German fighter, the co-pilot slumped over the radar set and the pilot's head lolling back, then it tumbled from the sky, spiralling down trailing a pillar of flame behind it.

One more threat remained, for the Me 109 had now extinguished its navigation lights. McLean tried to bring his guns to bear, but he could not get a bead on it and had to tell Evans, 'This one's all yours!' Evans's fire was accurate as the Me 109 made its approach, and McLean saw the muzzle flashes from the fighter's guns suddenly cease. It disappeared beneath the Lancaster, but McLean tracked it and unleashed a few more bursts into it as it emerged on the starboard side. The fighter disappeared from sight and several minutes later, having scoured the sky on all sides, McLean gave his pilot the all clear. 'That seems to be it, skipper. We're all alone again!' Only then, as the adrenalin from the fight began to fade, did McLean realise that he had been hit in the hand by one of the bullets from the first fighter attack.

As they made their way home, the wireless operator sent a signal to Group headquarters: 'Attacked by fighters east of Paris. Three twin-engined fighters definitely destroyed. One single-engined fighter possibly damaged. Lancaster slight damage, nothing serious. ETA base follows later, God willing.'

It is safe to assume that the message was not believed at first. Air gunners might register an occasional 'kill' against enemy night-fighters, and once in a blue moon they might even manage two, but to shoot down three, let alone four, was simply inconceivable. However, they landed safely back at base and the debrief with the intelligence officer confirmed the truth of their claims.

After a brief spell in hospital, McLean returned to the squadron and flew on a total of fifty-one operations before being rested. He had nine kills to his credit, and his prowess was recognised by the award of a DFC to go with his DFM.

★ ★ ★

The pace of 617's activities was continuing to increase, and on 16 March 1944, the night following Tom McLean's heroics on the abortive Woippy raid, the squadron was back in action as it 'celebrated Saint Patrick's Day' by flying to Clermont-Ferrand to bomb the Michelin rubber factory. They were accompanied on the raid by six aircraft from 106 Squadron dropping parachute flares over the target to illuminate it for the target markers: Cheshire, Munro, McCarthy and Shannon. Despite heavy fire from the anti-aircraft defences – 'there were so many shooting at us,' Cheshire's gunner, Wilfred Bickley, said, 'we went around again and dropped the marker' – all four of them put their red spot fires directly onto the target.

Reconnaissance photos later showed that all the Lancasters had scored direct hits. Ten 12,000-pound bombs and several sticks of incendiaries struck the factory site and every building apart from the canteen was destroyed or severely damaged. Cheshire 'hung around for about twenty minutes afterwards. The smoke was up about eighteen thousand feet. You could even smell it [the burning rubber]. We could see it for miles going back.'[23]

According to the intelligence officers who briefed them before the raid, the Maquis – the French Resistance – had told the factory owner to sabotage his own assembly lines, but he had not done so, although he had burned some of his stock in an attempt to pacify them. As a result they had labelled him a 'collaborator' with the Nazis. In the wake of the raid, the intelligence officers reported that the Maquis were now finding factory owners 'much more cooperative with them about sabotaging their factories'. They also claimed that the Gestapo was offering a $1,000 reward for anyone turning in a member of 617 Squadron.[24]

Shortly after the raid on the Michelin factory, Nick Knilans' crew were granted a nine-day leave. Knilans himself, fully justifying his reputation as a ladies' man, spent his leave in York with a WAAF officer from the Intelligence Section, even though he had previously been 'romancing a WAAF Corporal from her section. This officer had the Corporal posted off to another Station, unbeknownst to me. When I found out, I terminated our attachment. Besides, I had met a charming lass in the village ...'

He completed his first tour at the end of March and then faced a dilemma. He could have opted for a long rest period in Training Command, but would then have been sent back to America, while his crew would have been scattered and liable for a recall for a second tour of fifteen ops. Since 'Training Command would mean much more discipline and work', and they all liked living at Woodhall Spa, had many local friends, and 'the tripe had not been too dangerous', Knilans told his crew that he would see them through their second tour too.[25]

Only time would tell if Knilans would be stretching his luck too far.

CHAPTER 5

Spring 1944

Lancaster ED825

The spring of 1944 heralded a black day for Bomber Command when, on the night of 30 March, 795 aircraft were dispatched to attack Nuremberg, the iconic site of Hitler's pre-war rallies and the eponymous birthplace of his anti-Semitic laws. Nearly 100 aircraft and 700 men did not return from the raid. It would prove to be the RAF's deadliest night of the war: more men from Bomber Command died on that single night than the total aircrew losses during the entirety of the Battle of Britain. The disastrous operation would mark an end to the major attacks on cities and a concentration on preparations for the long-awaited invasion.

A few days later, on 5 April, Leonard Cheshire and the other Flight Commanders of 617 Squadron began marking targets, not just for their own squadron, but for the whole of 5 Group, a total of 144 Lancasters in all, ousting the Pathfinders who had previously done the marking. Whereas the Pathfinders dropped their markers from relatively high altitude, 617's crews used the low-level techniques that Martin and Cheshire had perfected. Their first target was an aircraft factory and repair workshops on the outskirts of Toulouse. Cheshire had previously used one of the squadron's Lancasters for his low-level marking, but he had now switched to a far faster and more manoeuvrable twin-engined Mosquito bomber. He would mark the aiming point with red spot fires, 617 would bomb those markers with a mixture of high explosive and incendiaries, and the other squadrons would then bomb on the fires that they had started.

The plan was supposed to be top secret, but it became clear that security had been breached when an Australian crewman from another squadron asked Nick Knilans about it. When he passed that information on to his Flight Commander, Les Munro, a major flap erupted. Knilans told his commanders that he knew the man was an Aussie based at Scampton but did not know his name. 'I did,' he later said, 'but why get him into trouble? Someone would get court-martialled. Somebody who had been in combat many times would have their life ruined. I suggested that they give a talk on security to all aircrew at Woodhall Spa and at Scampton. They did so.'

After a four-hour flight to the target, the Lancaster crews watched as Cheshire made two preliminary low-level passes over the factory roofs, once more giving the civilian workers a warning and the opportunity to take cover. 'I could see his Mosquito zooming about through a shower of light flak shells and searchlights,' Nick Knilans recalled. Cheshire dropped his markers on the aiming point, the aircraft repair hangar, on the third pass, and did so with such accuracy that the back-up markers, Les Munro and Joe McCarthy, did not need to mark the target at all. The first stick of bombs was a direct hit and the subsequent bombing almost obliterated the factory.

Don Cheney was making his 'second dickey' familiarisation flight on the attack – a trip all new pilots made as an observer with an experienced crew before embarking on operations. Watching

Cheshire marking the target and the 617 crews bombing the markers made him realise 'just how different ops were on 617 Squadron to Main Force; it was precise and exact, and not just dropping bombs all over the place.' The dangers of the latter approach were vividly illustrated later that week when inaccurate bombing during Main Force raids on French and Belgian railway yards saw 1,000 civilians killed.

While the Toulouse raid had been Don Cheney's first op with 617, it proved to be Johnny Johnson's last. He had already flown on twenty operations with the squadron, but his wife, Gwyn, was now pregnant, and when they landed after the attack on the aircraft factory, his pilot, Joe McCarthy, took him aside. 'Johnny,' he said. 'Gwyn doesn't know whether this child's going to have a father or whether she's going to have a husband. She must be worried stiff. You've got to give her a chance, pack it up now.'

His words made Johnson realise that although he didn't want to leave his crew, he had other, greater responsibilities. Gwyn had never expressed any worries to him or asked him to stop flying combat ops, 'but she was relieved when I told her,' he says. 'Though she said, "Are you sure?" Well I wasn't entirely, but I knew it was the right thing to do.'[1]

★   ★   ★

If Bomber Harris harboured any doubts about 617's low-level marking technique, they had now been removed, and he authorised 5 Group to continue operations completely independently of the Pathfinders and all other groups. Their first taste of marking for Main Force came on 18 April 1944, with an attack on Juvisy, just to the south of Paris, one of the largest railway marshalling yards in the region. In the months leading up to D-Day it was one of many sites targeted with the aim of disrupting German attempts to send reinforcements to Normandy once the invasion had taken place. That message was reinforced by 617's Station Commander, who told them that the success of any invasion would depend on the destruction of this and similar targets. If they were destroyed, it would not be possible for the Germans to mass additional men and munitions to counter the invasion when it came.

The marshalling yard was within a densely built-up area, with houses as close as 20 yards from the perimeter fence, and at the briefing the aircrews were warned once again of the danger to French civilians from misdirected bombs. The weather was good, with a fresh wind that would help to blow away the smoke and dust from the explosions, giving good visibility for the following aircraft. Their Lancasters were each loaded with four 1,000-pound and four 500-pound bombs, 'for immediate effect, no delays, just to smash the place up as much as possible to ensure that nothing and no one was able to go through for some months,' says John Pryor.[2]

It was a bright, starlit night and, looking down, Pryor, a twenty-seven-year-old Essex farmer's son with a square jaw and a steely, slightly intimidating gaze, could see the Channel clearly as they flew south. Had this been a raid on Germany, their bombs would have been armed from the moment they crossed the coastline, but flying over occupied France, they were not armed until the last moment to minimise the risk of civilian casualties. Flares had already been dropped to illuminate the target, but Pryor was ordered to drop back-up flares for extra light to avoid any mistakes.

Wing Commander Cheshire was in position and told his crews that the TIs (target indicators) were well positioned and to 'just go in and bomb them'. There was a heavy responsibility on both the navigator and the bomb-aimer, to make sure that, despite the smoke and dust from the previous bombs, their own were dropping inside the marshalling yard and not on somebody's house. There were no German night-fighters in evidence, only anti-aircraft fire 'of which we took no notice'.

Having dropped his own bombs accurately, Pryor was ordered to circle in case he was needed to suppress anti-aircraft fire, and so had a grandstand view as the Main Force began bombing, whereupon 'all hell was let loose'. Through the dust and smoke, he could see that the marshalling yard had been 'ripped to pieces' and, despite the fresh wind, dust and debris were now obscuring the target, so the Pathfinder Force began dropping parachute flares to guide the Main Force bombing. These drifted off in the crosswind and, though Main Force's initial bombing was on or close to the target, the later aircraft were dropping bombs as much as a mile to the north and west of the target, missing the yard and bombing the built-up areas around it. His aircrews could hear Cheshire on the radio, telling the Main Force that their bombs were missing the target, but it did not seem to have any effect on them. As he turned for home, still with another 200 or so Main Force aircraft queuing up to bomb, Pryor had 'a sinking feeling for the French people living in the built-up area that may have been hurt or even killed.'

As usual, Leonard Cheshire was the last to leave the target area and reported that the marshalling yard had been so badly damaged that 'not a square yard could be used for many a day'. That was confirmed by reconnaissance photographs showing that, while bombing had also extended for up to a mile to the north, the yard itself had been devastated. French civilians later confirmed that there was so much debris, smashed and broken trucks and trains, that it was 'long after D-Day' before it was possible to get even one railway line open again. The raid had been a great success, but between 100 and 200 people had been killed in the houses adjoining the yard, many as delay-fused bombs detonated while rescuers were trying to reach those trapped by the initial bombing. 'It mattered,' John Pryor said, 'but of course it had to be done.'

★   ★   ★

Although the success of the low-level, precision target marking for 617 Squadron instigated by Mick Martin and Leonard Cheshire had led to it being 'rolled out' for 5 Group too, there were still powerful voices within Bomber Command who claimed that, while the technique might be effective against targets in occupied France, it would be 'suicidal' to adopt low-level marking against heavily defended targets in Germany. The ultimate test-bed for its wider application therefore came on the night of 24/25 April 1944, when Cheshire led four Mosquitos to mark targets in Munich for a large force of 5 Group bombers. It was made clear to Cheshire that the allocation of the Mosquitos to 617 Squadron would only be made permanent after an effective demonstration of their effectiveness in low-level marking over the target.

The new technique had already been used at Brunswick two nights before the Munich raid, but due to the radio of one aircraft being stuck on 'transmit', drowning out all other radio traffic, the attack had been a disaster. Having failed to achieve the required results at Brunswick, the Munich raid was now seen as make or break for the low-level marking technique.

If Cheshire and the other pilots revelled in the speed and manoeuvrability of the Mosquito after the ponderous Lancaster, their navigators were less enamoured. Conditions for them were 'almost viciously cramped compared to the Lancaster'.[3] There was no navigation table and the navigator had to sit with a board across his lap and the charts and log sheet pinned to it. Gerry Fawke's navigator, Tom Bennett, found that 'Operational necessity demanded that the dividers, Dalton computer and India rubber be firmly tied by appropriate lengths of string to the clips of the parachute harness if they were to remain available at all times, despite the gyrations of the aircraft in the close operational role.' He kept his pencils inside the forehead covering of his helmet and his Perspex ruler pushed down the leg of his right flying boot, in common with all Bomber Command navigators. 'I always figured that the Germans must have had great stocks of these Perspex rulers, taken from shot-down navigators!'

In addition to directing his pilot to the target, the navigator performed several other tasks. 'The pilot never moved from his seat. He sat there in lordly authority, directing the navigator with an imperious wave of his hand when he wanted petrol tanks changed, or to crawl forward into the nose, to fuse and select the markers ... At times the navigators of the Mosquito Marker Force were busier than one-armed paper-hangers!'

As an added stress on the Munich flight, even starting from RAF Manston in Kent, the Mosquito crews could expect to arrive back at Manston with 'a spoonful of petrol' – no more than ten minutes' fuel in their tanks, compared with the normal safe margin of one to two hours'. However, pilots and navigators considered the odds and accepted them, and there was even 'a quickening of adrenalin flow at the thought of a challenge to their professionalism'.

At the final briefing, the normal banter between crews was in full flow, the Lancaster navigators helpfully advising the Mosquito crews that the way to find Manston was to fly south until they hit the coast and then turn to port and follow the coastline all the way to the North Foreland, by which time Manston should be in sight. The Mosquito navigators accepted that advice in the spirit in which it was offered!

After arriving at Manston, the four aircraft were left as close as possible to the take-off point and the fuel tanks were then filled to the brim. Even so, the navigators calculated that they would have no more than six minutes over the target if they were to return before their fuel ran out. Such was the speed of the Mosquitos, compared with the ponderous Lancasters, that when they took off at one minute before midnight on 24 April 1944, it was three hours after the 244 Main Force bombers had departed.

There was ten-tenths cloud over northern Germany, but the skies had cleared by the time they reached Munich. The marking force ahead of them was already dropping flares to illuminate the target area and Mosquitos from 627 Squadron were also dropping Window to confuse the German radar. Leonard Cheshire directed the flare force and then called 'Tally ho!' as he sighted the aiming point and put his Mosquito into a dive. Munich was defended by over 200 light flak guns, which put up a barrage of shells, but Cheshire dived through the clouds of flak bursts, dropped his markers at 300 feet and then called in the other three to drop their markers onto the spot fires that he had laid. According to Bob Knights, they marked it with such accuracy that 'There was this magnificent aiming point of a circle of spot fires, which they couldn't miss.'[4]

Wilfred Bickley also saw Cheshire put his marker 'right in the middle of the town', and after the op asked Cheshire's navigator, 'What was it like?'

'I don't know,' the navigator said. 'I had my bloody eyes shut.'[5]

The 617 Squadron Lancasters were first to bomb, clustering their high-explosive and incendiary bomb-loads closely around the spot fires, both beginning the destruction of the target and acting as even more unmistakable markers for the succeeding squadrons, and as the Mosquitos turned for home, Tom Bennett 'saw the agony of Munich already beginning', with bombs raining down on the Bavarian capital. All four Mosquitos were coned by searchlights – highlighted by multiple beams – but successfully evaded them. Dave Shannon was coned at the precise moment when he'd asked his navigator, Len Sumpter, what course to set. 'Drop your starboard wing and I'll tell you,' Sumpter said. As he did so, Sumpter held up his chart and used the glare of the searchlight beam to read off the new course for his pilot to follow, 'a first-class example of using the resources of the enemy for one's own benefit!'[6]

All four aircraft landed safely back at Manston, though not without incident. As Cheshire was coming in to land, he was warned by the ground controller that an enemy intruder was circling the airfield. The runway landing lights were then switched off. Cheshire told his navigator to turn off their navigation lights to make them a more difficult target for the raider, but in his haste, the navigator turned off the whole electrical system, including the power for the navigation kit and the fuel pumps, and hastily had to remedy his mistake. The landing lights were still off when Gerry Fawke came in to land, which he mistook for 'electrical trouble with the airfield lighting',[7] and he was also fired on by the intruder. Fortunately his aircraft was undamaged, but the lack of runway lights caused him to overshoot and he came to a halt with the undercarriage buried in the sand dunes beyond the end of the runway.

The delight and relief of the Intelligence Section staff at the successful return of the Mosquito crews showed that they had been aware of the gamble that they had taken. 'The post-op mug of hot sweet tea was always to be savoured by returning aircrew, but when it was served up by a pretty young WAAF whose demeanour showed how overjoyed she was at our safe return, and when it was accompanied by some very special (for wartime) sandwiches and biscuits, then I must admit that one did feel the effort had been appreciated.'[8]

When Gerry Fawke's aircraft was extracted, undamaged, from the sand, he discovered that their entire bomb-load of markers was still in the bomb-bay. Fawke spoke for his navigator too, when he said, 'God, Ben, all that chase down from Woodhall, economical fuel attention all the bloody way to Munich and back, shot at by a Jerry night-fighter, an argument with a heap of sand and the bloody markers never left the aircraft. What a complete waste of time and effort!'

However, if it had been a waste of time for Fawke's crew, the raid on Munich had been spectacularly successful in terms of bombing accuracy, destroying well over 1,000 city-centre build-ings, and giving a vivid demonstration of the effectiveness of low-level marking. As a result, 617 Squadron retained its four Mosquitos, and 5 Group – and later Bomber Command's entire Main Force – duly adopted the marking techniques developed by Cheshire's squadron. Three Pathfinder squadrons were perma-nently transferred to 5 Group, which became known by other squadrons, in what was partly disparagement and partly a grudg-ing compliment, as 'The Independent Air Force'.

Having not slept for twenty-four hours, the Mosquito crews flew back to Woodhall Spa early the next morning. Aircrews often took 'wakey-wakey pills' – Benzedrine tablets – to keep them awake, but they had taken none on this trip. Navigator Tom Bennett was daydreaming, reviewing the previous night's op as the aircraft cruised at 3,000 feet, when the combination of his fatigue and the warm sunshine caused him to drop off for a few moments. He awoke with a start, berating himself for having dozed off, but as he glanced at the altimeter, he saw that they were down to 700 feet and still in a shallow dive. When he looked at the pilot, he discovered that he too had 'succumbed to the blandishments of Morpheus' and was fast asleep with his chin on his chest. With commendable alacrity, Bennett released his seat harness and leaned across to ease the control column back with his right hand, while simultaneously thumping Fawke on the shoulder with his left and bellowing 'WAKE UP, MATE!' into his ear. When Fawke had regained the aircraft's altitude and attitude, he gave Bennett an embarrassed smile and a wink. Strangely, the incident slipped both their minds when completing their post-op debriefing ...

All the 617 Squadron aircrews had been granted seven days' leave, and the dog-tired Bennett then managed to fall asleep again both in the car – fortunately driven by someone else – taking him to Grantham station, and then on the train to King's Cross. He awoke as the train came to a halt in London to find that he had 'slumbered gently throughout the journey with my head cushioned by the soft shoulder of an attractive girl'. He stammered out an apology but she brushed it aside, saying that she would 'consider it part of her contribution to the war effort'!

CHAPTER 6

The End of the Beginning

In late April 1944, as part of the build-up towards D-Day, Exercise Tiger was taking place off the south coast of Devon. It involved 30,000 American troops, who were to practise a beach landing at Slapton Sands under live fire. A communications failure first led to a 'friendly fire' incident in which American troops were shelled by a British cruiser, but worse was to follow. A British corvette and a line of nine landing ships laden with American troops came under fire from German E-boats which had evaded British naval defensive screens. Two landing craft were sunk and two more badly damaged, with the loss of 750 US servicemen. Official embarrassment over the disaster, coupled with the need for secrecy ahead of the D-Day landings, meant that all information about the disaster was suppressed and its full scale was not acknowledged until after the war was over.

617 Squadron's own preparations for D-Day began a few days later. At eleven o'clock on the morning of 2 May 1944,[1] the crews were summoned by tannoy to a briefing, and told to bring their identity cards with them, something which had never been required of them before. When they reached the briefing room they found the door guarded by Military Police (MPs), who scrutinised each ID card with microscopic attention before allowing them to pass. Inside, there was no sign of the WAAFs who normally produced the maps and intelligence data for the ops, nor the officers of the Intelligence Section. The telephone exchange had been shut down and the control tower cleared of all personnel. More MPs formed a cordon around the building.

Meanwhile, Air Vice Marshal Cochrane was pacing along the perimeter track with 617's commanding officer, Leonard Cheshire, and had only one question for him: could he vouch for the ability of every single member of the squadron to keep silent about an absolutely vital secret for a period of at least six weeks? After a few moments' thought, Cheshire answered 'Yes' and the two men retraced their steps to the briefing room. To help ensure that Cheshire's men justified his faith in them, agents of the Military Police's Special Bureau of Investigation were 'very active' in the Lincolnshire area over the next few weeks, alert for any whisper that could be traced back to 617 Squadron, but not a hint of any leak was discovered.

Cochrane's opening statement was electrifying: 'Gentlemen, the next time you are airborne it will be D-Day!'[2] The aircrew leapt to their feet as one man, cheering and pounding the tables with their fists. It was several minutes before the hubbub had died down enough for Cochrane to resume his speech. He told them that the War Cabinet wanted aircraft using Gee navigation (the radio-based system that measured the time delay between two radio signals to produce an accurate positional 'fix') and Window to create the impression of an invasion convoy on the screens of German coastal radar stations. Cochrane told them:

> *The problem was first referred to the Fleet Air Arm, as they have the slowest aircraft available but they have said that it is beyond their scope. Next it was offered to Coastal Command, since the operation will, of necessity, be in their special sphere of operations. Coastal have declined on the grounds that it is not possible to simulate an eight-knot invasion convoy with aircraft travelling at over 100 knots. The problem was then put to Bomber Command. Sir Arthur Harris took one look at it and said, 'Send this to 617 to solve and let the rest of the Command get on with our business of fighting the war!'*

'So gentlemen,' Cochrane said, after pausing to assess the impact of his words:

I'm the bearer of this message. 617 Squadron is required to simulate an eight-knot naval convoy using your Lancasters cruising at 160 knots. There is no outside help available to you this time, absolutely none. You are on your own with this. There is not even a suggestion of how to solve it. On the face of it, it is impossible, but I share the Command Staff's faith in the squadron that it will be solved, that 617 will rise to this challenge as they have risen to all others. Of one thing I am certain. If you, with all your operational experience and flair, cannot come up with a solution, then it is truly insoluble!

No aircrew in this room is permitted to engage in operations until after D-Day, whenever that may be. I will not harp on the security angle too much, for the consequences of such a leak are too horrific to contemplate. All of you are responsible for each other. A failure by one single member will be counted as a failure by the whole squadron.

The squadron's navigators immediately put their heads together and began wrestling with the apparently impossible problem of reconciling the speed difference between the Lancasters and the snail-like pace of the surface ships. They eventually air-tested a number of options over the following days until they had arrived at what they felt was a workable solution: flying oval circuits above a small surface force as it advanced towards the French coastline while dropping Window to simulate a mass fleet approaching on the enemy's radar screens. They then requested 'boffin help' to refine the system and determine the optimum height to fly and the quantities and frequency of Window drop. A distinguished physicist, Dr Robert Cockburn, the leader of the scientific team that had originally developed Window to defeat German radar, worked with the squadron's navigators to refine and perfect the system.

Cockburn had joined the Royal Aircraft Establishment, Farnborough, in 1937 and worked on the ground-to-air VHF communications that were of vital importance for RAF Fighter Command during the Battle of Britain. In 1940 he was assigned to the Telecommunications Research Establishment near Swanage, where he headed a team working on radio counter-measures – 'The Battle of the Beams' – when successful jamming of the German bombers' navigation systems reduced the damage they caused to London during the Blitz. Cockburn's team also developed devices to jam enemy radar, reducing losses suffered by RAF bombers over Germany.

Window, dropped by aircraft, created a radar echo similar to that of an actual bomber, confusing the air defences. It proved so effective in testing that some senior RAF officers opposed its introduction because of the potentially disastrous consequences if the Germans discovered the secret and turned the invention against its British creators. Bomber Command was finally permitted to use Window for the first time during Operation Gomorrah over Hamburg and it led to an immediate reduction in aircrew losses.

After Cockburn and the 617 Squadron navigators had revised their system, it was tested against a captured German radar installation set up in the Firth of Forth, and on the Yorkshire coast near Flamborough Head, so that the radar picture could be observed while 617 Squadron flew their prescribed patterns offshore. It was not something most of the aircrew enjoyed doing. 'It was frustrating just to be flying around the North Sea,' John Bell says, 'carrying out navigation exercises, turning this way and that, as we were observed on radar from Flamborough Head. I don't think we even dropped Window until the last exercise. We just thought we should be dropping bombs on targets.'

Those not taking part in the day's drills often flew up to Driffield and then went by road to the radar station at Bempton to see the deception unfold on screen. A few of the more adventurous members of the squadron also spent some of their spare time climbing down the sea cliffs at Flamborough to look for seagulls' eggs which were then 'brought back for tea'.

John Bell may have been bored with the invasion rehearsals, but life, and death, continued apace for the vast majority of the men of Bomber Command.

⋆   ⋆   ⋆

Cochrane's prohibition of 617 aircrew taking part in operations until after D-Day was lifted at once when agents in France reported that the German 21st Panzer Division of some 15,000 men had moved into the barracks at Mailly-le-Camp, the biggest army camp in north-west France. The absolute necessity of destroying the Panzer division to prevent it intervening against the forthcoming D-Day invasion led to permission being given, albeit with considerable reluctance, for the four 617 Mosquitos, led by Cheshire himself, to mark the target for 346 Main Force bombers. However, the division of responsibilities between Cheshire, acting as lead target marker, and Wing Commander Laurence Deane of 83 Squadron, who was the overall controller, was a recipe for potential confusion, if not disaster.

Rusty Waughman

The CO of 166 Squadron had assured his men that the op would be 'a piece of cake, chaps, just like falling off a log. Tonight there will be no night-fighters, very little ack-ack, just go in and wipe it off the map and come home.'[3] Flight Lieutenant Russell 'Rusty' Waughman, a pilot with 101 Squadron, also recalled thinking at his briefing that 'this was going to be one of the easier operations', but the crews were all to be rapidly disabused of that idea.[4] One crewman from 166 Squadron stopped for a word with one of his mates from a different crew as they were leaving the briefing. His friend told him that 'He'd had a rotten leave, he had fallen out with his girlfriend,' and didn't like the operation as it was out of Gee range, and 'without it his navigator was useless. His final words were, "This operation, we've had it." We had a good laugh and parted. He was lost with the rest of the crew that night.'[5]

The yellow TIs – target indicators – at the assembly point were laid late, and the Main Force bombers were then held there. 'We did not like this at all,' one crewman recalled:

and the crew were worried as visibility was clear and we knew from experience the dangers of hanging around enemy territory any longer than absolutely necessary. We were circling this flare for approximately half an hour and becoming increasingly worried as it appeared impossible to receive any radio instructions due to an American Forces Broadcasting Station blasting away. I remember only too well the tune, 'Deep in the Heart of Texas', followed by hand-clapping and a noise like a party going on. Other garbled talk was in the background but drowned by the music.[6]

That made it 'very difficult, if not impossible, for the Main Force to hear any instructions from the M/C [Master of Ceremonies],' recalls Rusty Waughman.

Having marked the target, Cheshire tried to call the Main Force bombers in to attack, but Deane did not respond and the interference meant that Cheshire went unheard by the Main Force crews. Since they received no orders to begin their bomb-runs from Deane either, they were forced to keep circling to the north of the target, the bright moonlight making them sitting ducks for the flak batteries and the prowling night-fighters. One crewman, whose job was to stand in the astrodome to report sightings of aircraft being hit and record the positions according to a clock dial, suddenly saw 'a burst of green tracer fire flash across the sky, then another, and so it went on and there were aircraft going down in all directions.' He counted seventeen bombers shot down before the navigator told him to stop.[7]

'Things hotted up quickly,' Waughman said.

There was a lot of German fighter activity and Main Force aircraft were
seen being shot down. The R/T discipline, I'm afraid to say, was bad.
There were many skippers calling for the okay to go in and bomb; their
language was fruity to say the least, the night sky was blue! One pilot was
heard to say that he was on fire and for the markers to 'pull their fingers
out'. An Australian voice came in reply: 'If you are going to die, die like a
man: quietly.'

Some Main Force crews took it on themselves to begin their
bomb-runs, but the rest had to wait until 00.24, when they
finally heard the order 'Panthers, go in and bomb!' By then the
flak and night-fighters had already taken a very heavy toll of the
bombers. 'All too often we would see flames above and soon
afterwards a heavy bomber would fall to the ground,' one pilot
recalled.[8] And by the time one crew began their run at 00.30,
they could already see 'several planes burning on the ground'.[9]
Another crewman watched in horror as 'two of our companion
planes burst into flames. We managed to pick out a tiny figure
escaping from his inferno, looking like a little toy parachutist
suspended in midair.'[10]

Rusty Waughman's crew saw another aircraft blow up immediately below them. The bomb-aimer, who was lying down in
the nose, saw the explosion and the blast and flames flaring rapidly towards him but had no time to say anything before the
blast wave hit them. 'We were blown nearly completely upside
down,' Rusty said.

I recall having to pull back very hard on the control column, as if coming
out of a very badly executed loop the loop. The airspeed indicator read
nearly 400 miles an hour [the Lancaster's maximum speed was supposed
to be 282 miles an hour]. This happened at about 7,000 feet; we eventu-
ally pulled out at just under 1,500 feet, heading down very fast, with the
aircraft creaking under the exertions. This happened just after we dropped
our bombs; I doubt if we would have recovered if we'd still had our load
on board.

Despite the carnage among them, Main Force bombed the target very accurately and 'brilliant yellow fires were seen all around the A/P [aiming point] and the smoke billowed up to a great height. Shockwaves were seen, caused by the bursting bombs, radiating like ripples caused by a stone dropped in a pond.'[11]

In strictly operational terms, the raid on Mailly-le-Camp proved a great success. Eighty per cent of the tank depot had been destroyed, with heavy losses of tanks and other vehicles, and heavy casualties among the tank crews and other trained personnel. But Main Force had paid a savage price for the operation, with forty-two bombers shot down – one in eight of those which had taken part. Understandably furious at their losses, many Main Force personnel blamed Cheshire for the disaster, though this was completely unwarranted. Apart from the interference from the American radio station, Deane's radio was later found not to be tuned to the designated frequency, causing his signals to be weak and difficult to receive.[12]

'It was a night that I shall never forget,' one Main Force survivor recalled, 'and I think that a lot of boys were turned into men on the raid.' At the debriefing for 166 Squadron the CO demanded to know what had gone wrong. 'A Flight Sergeant who was on his third tour of operations – in the region of ninety operations including trips to Berlin – turned round on the Commanding Officer and told him to "(expletive deleted) off. And don't you talk to me about falling off a log. Give me Berlin any time." The Commanding Officer quickly disappeared.'[13]

The raid had one other consequence. Until Mailly-le-Camp, raids on French targets were regarded as so routine that they were counted as only one-third of an operation – had a crew flown only on ops over France, they would have required ninety ops to complete a tour, rather than the usual thirty. After Mailly-le-Camp, each raid over France counted as one op.

A senior intelligence officer, fearing that if shot down, one of the 617 Squadron crewmen might reveal the secrets of their D-Day deception under Gestapo torture, had had some kind words for them as they had prepared for the Mailly-le-Camp op:

'Gentlemen, if any of you go missing, it will be a relief to us to hear that you are dead.'[14] However, to his – and their – huge relief, and despite the huge Main Force losses, all four crews of 617 Squadron returned safely to base.

★   ★   ★

For the rest of 617 Squadron, the tedium of the endless rehearsals for Operation Taxable – as their contribution to the D-Day landings was known – and no 'live' ops to relieve the monotony, was only partially relieved by low-level 'beat-ups' of the Norfolk coast as they returned to base. The coastal fishermen, bait diggers and potato pickers startled out of their wits by 617's Lancasters blasting in at wave-top height may not have been fully appreciative of the pilots' need to let off steam!

On 16 May there was a genuine reason for high spirits as the squadron celebrated the anniversary of the Dams raid. Many veterans of the raid joined a celebration party at Petwood Hall. It began with a 'somewhat formal gathering' on the communal site in front of the NAAFI, with officers and airmen together, and became considerably less formal as the day wore on. It's likely that sore heads meant the practice for Operation Taxable the next morning was not performed with quite the customary precision, though the worst hangovers came twenty-four hours later.

The Station Commander at Woodhall Spa had decided that since many VIPs and their wives were attending the party and 'due decorum should be maintained', a marquee should be erected on the lawn at the rear of the Petwood and several barrels of beer installed there so that the junior officers could 'drink to their hearts' content' without any risk of embarrassment to the VIPs. In the event, it poured with rain for the whole of the afternoon and evening, and nobody went outside at all. The next morning, to the junior officers' delight, they discovered six untouched twelve-gallon barrels of beer still sitting in the marquee. 'As we were not on ops for the next two nights,' one of them recalled with a smile, 'you can imagine the result!'[15]

Guy Gibson, the celebrated leader of the Dams raid, who had now returned from his lengthy flag-waving tour of the United States, appeared on *Desert Island Discs* and completed a spell at Staff College, had been unable to attend the party, or the impromptu junior officers' booze-up in the marquee that followed it. But he did attend the all-ranks Squadron Dance held later that week, on 19 May. There was a large trestle table at the front holding a huge cake, with icing-sugar replicas of the infamous dams decorating the top. Cheshire said a few words first and then Gibson climbed up on to the table next to the cake to give his speech. As he tried to get down afterwards, he slipped and sat down squarely on the cake. 'His backside was covered in icing and there was a huge amount of cheering and hooting!' John Bell says, smiling at the memory.

Don Cheney recalls another party at 106 Squadron when Guy Gibson came back for a birthday party thrown in his honour. 'I remember him serving at the bar wearing a corporal's dress jacket while the corporal steward wore Gibson's dress jacket, resplendent with all his medals, including the VC! He seemed a very pleasant guy and joined in all the fun ... I seem to remember at one point that someone brought a horse into the party!'[16]

Still carrying out endless practices with no idea how long they would have to continue, 617's crews chafed at the lack of action and welcomed any chance to alleviate the boredom. Shortly before D-Day, an opportunity presented itself when, fearing that German paratroopers might attack airfields to disrupt air support for the D-Day invasion, Leonard Cheshire issued his men with arms – pistols and Sten guns – for the defence of themselves and their base. Three days later, the arms were hurriedly recalled 'after much ribald target-shooting' had smashed scores of the hotel's china plates and there had also been a few near-misses by shots fired from the upper windows of the Petwood.[17] 617's aircrew may have been able to drop a bomb into a barrel from 18,000 feet but, issued with small arms, few of them seemed capable of hitting a barn door from twenty paces!

On 5 June 1944, the day after the Allies had entered Rome, the squadron's endless rehearsals came to an end with confirmation that D-Day – Operation Overlord, the invasion of Normandy – was set for the following day. If there was disappointment that 617 would not have a more active involvement in D-Day, the crews were assured that their role was nonetheless an absolutely crucial one, a message reinforced by Leonard Cheshire in their final briefing.

'The waiting is over,' he said:

Not just the period of the recent training through which we have all been, but also for the years we have fought our way to this day. This is possibly the most crucial operation this or any other squadron has ever been called upon to perform. Our efforts tonight will not be of the usual destructive nature, but our successful endeavours will undoubtedly save hundreds of Allied lives this night, and possibly thousands in the weeks to come ... Aircraft of the Second Tactical Air Force will sweep up the Seine soon after dawn tomorrow and break all the bridges between Le Havre and Paris, effectively isolating the large German armies in northern France, which our operation tonight is designed to hold in position in the Pas de Calais. I have not the slightest doubt about the successful outcome of tonight's sortie. Thank you gentlemen, and as always, the best of luck to you all![18]

'There was real excitement' at the news, John Bell says. 'The invasion was happening and we were part of it! Everyone was keyed up in preparation and raring to go, but we had to get it right because thousands of lives depended on it. We all understood that at the time.'

Allied bombers had already been pounding gun batteries and radar stations along the entire French coast, trying to avoid revealing the true focus of the impending attack by inflicting equal damage on areas far from Normandy. A thousand aircraft were involved in D-Day operations that night, and the fuselages and wings of all Allied aircraft, including 617's, were painted with black and white stripes, like a piano keyboard, to minimise the risk of losses to 'friendly fire', since the volume of D-Day signals traffic was certain to swamp the normal IFF (Identification Friend or Foe) system that used a transponder to identify friendly aircraft to British radar. Bomber crews were also forbidden from jettisoning bombs over the Channel that night, because of the significant risk of hitting one of the hundreds of Allied ships making the crossing. To reinforce Operation Taxable's simulation of an invasion at close to the Channel's narrowest point, hundreds of fake aircraft, landing craft and military vehicles had been assembled in Kent for the benefit of German air reconnaissance and spies. All was now in readiness for the grand illusion to be created.

'Although it was the invasion, we were not thinking about the end of the war,' one of the 617 Squadron pilots later recalled. 'Just that this was a huge step forward in the course of the war, possibly a defining moment in what the future might hold. But we certainly weren't talking about "the end".'[19] Les Munro was the first 617 Squadron pilot airborne at 23.05, with Cheshire and the other six aircraft of the first wave just behind him. They rendezvoused with the surface craft just off the south coast. The eighteen ships – harbour defence motor launches and search-and-rescue pinnaces – towing barrage balloons behind them and broadcasting sound effects and using radar counter-measures to strengthen the deception, began steaming slowly towards France. Meanwhile, 617 Squadron's Lancasters flew overhead, travelling at a precise speed and height – 180 miles an hour and 3,000 feet – in line abreast with a separation of two miles between each aircraft. They held their course for exactly two and a half minutes, dropping their Window – the strips of aluminium chaff were 6 feet long and tied in bundles of 100 – at the rate of one bundle every four seconds.

They then performed a slow 'Rate One' turn lasting one minute, dropping no Window whilst turning, and then resumed dropping on the slightly shorter, two minutes and ten seconds leg back towards the English coast, before turning again, overlapping their previous track. Each completed circuit advanced them a little further towards France – one mile every six minutes and forty seconds – keeping pace with the ships below so that, seen through German radar, a huge convoy of ships appeared to be crossing the Channel towards Cap d'Antifer, just north of Le Havre, at a speed of 8 knots.

With a double crew and boxes and bundles of Window of different sizes stacked in every available space, each aircraft was very crowded. There was a red and green light operated by the navigator which told those dropping Window when to start pushing the bundles out and when to stop. When the next person took over, it had to be carefully coordinated so there was no pause or change in the rate the bundles were being dropped. Progressively larger bundles of Window were used as they flew nearer to the French coast, producing the increasingly strong radar signal that the Germans would have expected from an approaching fleet. All the time the navigator was keeping up a constant stream of instructions to the pilot: 'Tighten the turn, you're two seconds slow ... You're three feet too low. On course, on course. Begin to turn ... Now! Ease up, you're three seconds too fast.'[20]

Operation Taxable continued for four hours, and the work of dumping the Window down the flare chute was so gruelling that each Lancaster's double-sized crew – fourteen men – took turn and turn about. The intense concentration required of the pilots and navigators, who had to keep within four seconds elapsed time and five feet of altitude at all times, meant that they were also 'spelled' by replacements. As the first wave of eight Lancasters ran short of fuel and Window, a second wave of eight replaced them. German radar picked up the ghost fleet, and confirmation that they had bought the deception was offered as some of the shore batteries began firing on the non-existent invaders. In the final stage, the ships ran in fast to within a couple of miles of the shore, simulating a landing attempt, and then retreated under cover of smoke, laying mines behind them.

Just before the dawn light that would expose the hoax to the German gun batteries preparing to open a barrage of fire against the phantom invasion fleet, the Lancasters and their attendant ships headed for home. 617's men had to keep below 1,000 feet, since the skies above them were thick with outward-bound bombers, fighters, and gliders carrying assault troops. On the surface of the sea below them, they could also see that the Channel to the west was black with ships making for the true invasion sites on the Normandy beaches. 'As we headed back, D-Day was under way,' John Bell says, 'and there was a real sense of a momentous day and great excitement – if the invasion was happening, then we really were on the front foot. But we still had no idea how long the war would go on for. Not a clue.'

Several other simultaneous deception exercises had also taken place that night. Operation Glimmer was a similar Window-dropping exercise carried out by 218 Squadron and six motor launches, simulating an invasion fleet approaching Boulogne. Operation Moonshine involved a small flotilla of boats a few miles off the French coast, deploying radar-reflective balloons. They were linked to a device that amplified German radar pulses, making the flotilla appear on enemy radar screens as a large fleet. Loudspeakers on board the small craft broadcast the sound of ships dropping anchor to strengthen the deception.

Meanwhile, in Operation Titanic, forty aircraft dropped hundreds of dummy parachutists: 200 south-west of Caen and another 200 south-west of Dieppe. The dummies were cloth bags fitted to a cross-shaped wooden frame with an explosive charge that destroyed the dummies on landing, making it look as if the parachutists had burned their chutes and were now in hiding, ready to carry out ambushes or sabotage.

The aircraft also dropped rifle-fire simulators, bundles of Window and, alongside the dummy parachutists, two genuine SAS teams of paratroopers to reinforce the illusion that airborne landings were taking place. The SAS men had orders to carry out sabotage and take prisoners, but they were also instructed to allow some of their captives to escape, so that their reports to their German commanders would strengthen the impression of an invasion near Dieppe, rather than the true sites in Normandy.

While all this was taking place, Operation Mandrel saw aircraft armed with radar jammers circling over the Channel between

Littlehampton and Portland Bill, creating a jamming screen to hide the real invasion fleet from German radar, while specialised Lancasters and Flying Fortresses carried out radio-jamming and dropped a Window barrage over the Somme estuary to draw enemy night-fighters away from the vulnerable transport aircraft carrying airborne troops towards Normandy.

Operation Taxable and the other deceptions proved to be a huge success. Apart from the 'convoy' being fired on by shore batteries, intelligence intercepts showed that German commanders in the region had reported an approaching invasion fleet. The German forces had been fixed in place in the Pas de Calais and the destruction of the Seine bridges by British and American bombers later that morning left them trapped, unable to reinforce the embattled German defenders in Normandy. Well aware of the extraordinary efforts his squadron's navigators had made to find a solution to an apparently intractable problem, Cheshire recommended several of them for decorations, but all were turned down with the terse comment, 'This operation did not cross the enemy coasts, or at any time come under enemy gunfire. Consequently, it is not eligible to be considered for the award of decorations.'[21]

D-Day may have marked, in Churchill's words, 'the end of the beginning', but there were still many perilous months ahead, and many further occasions where 617 Squadron would demonstrate their courage, and their sacrifice.

CHAPTER 7

The Fight Goes On

In the weeks following the invasion, Bomber Command's efforts were primarily directed in support of the advancing Allies and the tempo of operations was ramped up by both day and night. During this period the command flew the same number of sorties in an average week – over 5,000 – as had been flown in the first nine months of the war. Between June and August over 180,000 tons of bombs were dropped, with only 30,000 targeted against Germany.[1]

On 9 June 1944, 617 Squadron was tasked with blocking the Saumur tunnel 125 miles south of the main battle area, on high ground above the river Loire, to stop a crack Panzer division that was being brought north by train to attack the Allied troops as they consolidated their Normandy bridgeheads. The raid, planned in great haste, would be the first operational use of a new 12,000-pound bomb developed by Barnes Wallis in conjunction with Vickers Armstrong.

Back in April, while the rest of the squadron had been bombing the railway marshalling yards at Juvisy, south of Paris, Nick Knilans' crew, sworn to absolute secrecy, had been sent on a special duty to test a new weapon. Although the idea of 12,000-pound bombs was not new, the previous versions had been little more than three 4,000-pound 'cookies' bolted together. The new bomb, christened 'Tallboy', was custom-designed. Twenty-one feet long, it was torpedo-shaped and highly polished, with a slim, streamlined design that belied its 12,000-pound weight. It had a long tail of light alloy, four small square fins that were offset by 5 degrees to impart spin as the bomb dropped, greatly improving its accuracy, and a hardened, cast-steel nose to withstand the impact as it struck the ground and increase its penetration below the surface.

Dropped from an optimum height of 18,000 feet at an airspeed of around 170 miles an hour, a Tallboy would take 37 seconds to fall to earth. By the time it hit the target, it would be travelling at a terminal speed of about 1,100 feet per second – 750 miles per hour, roughly the speed of sound. Its slim profile, hardened nose and huge kinetic energy allowed it to drill down deep into the earth before exploding, causing a seismic underground 'earthquake effect' that meant a near-miss could be as devastating to a target as a direct hit.

A Tallboy 12,000-pound bomb falls earthward

Having loaded the Tallboy into a Lancaster that had been specially modified to carry it, Knilans' crew flew over the bombing range at Ashley Walk in Hampshire and dropped the bomb from 20,000 feet. A camera had been placed right in the centre of the target to record the bomb in flight and the ensuing explosion. 'The idea would have been a good one if Joe [the bomb-aimer] had been off target. He was bang-on again. The target became a crater, eight feet deep and a hundred feet across. The camera was never seen again.'

Wallis assured Sir Arthur 'Bomber' Harris that his new bombs could penetrate 15 feet of reinforced concrete. The Allies had no other weapon that could inflict such damage on the concrete blockhouses the Nazis were building to protect their new V-2 'terror weapons', and only one squadron – 617 – could be relied on to drop them with sufficient accuracy from 18,000 feet, at which height a blockhouse would look no bigger than a pinhead.

The Tallboys had been delivered to Woodhall Spa in crates marked 'kitchen boilers' and were 'hiding in the bomb dump', so that even Leonard Cheshire did not discover 'what these massive bombs were' until shortly before the raid on the Saumur tunnel.[2] When he eventually saw the new weapons, Les Munro remembers thinking, 'Hello, this is going to be good!'

The squadron's ground crews weren't used to loading the new bombs, and they had to make rapid alterations to the bomb-bays of the Lancasters to house them,[3] but loading them was still a complicated, time-consuming business. Three fuses had to be fitted into the rear of the nose section before the tail unit was bolted into place, and the bomb then had to be lifted by a six-ton crane onto a custom-built cradle on a trolley. The trolley was then towed by a tractor underneath the aircraft, and such was the bomb's size and weight that, rather than winching it up into the aircraft's bomb-bay, four armourers standing at each corner of the cradle had simultaneously to jack up the bomb into position. A metal band around the middle of the bomb secured it in place and arming wires were connected, allowing it to be fused in flight.

Three Mosquitos and twenty-five Lancasters of 617 Squadron took part in this first Tallboy raid. In pitch darkness, Cheshire flew in at 200 miles an hour, 50 feet above the ground, straight along the railway tracks towards the southern mouth of the tunnel. At the last moment he dropped his red spot fires, then pulled up in a steep climb and radioed to his men, 'OK, A Force. Start bombing red flares. They're right in the mouth of the tunnel.'

When the last Tallboy had been dropped by the first group, Cheshire flew back across the tunnel mouth to check the results of the bombing and then called in B Force to finish the job.

Mac Hamilton's crew had been the eighth in line for take-off, but they were still trying to get the bomb adjusted on their aircraft when the others took off. An hour had passed by the time the problem was solved, but when Mac told his crew that it was too late for them to join the raid, his navigator said, 'Not if we go straight there,' because the others had flown a circuitous route to keep the air defences guessing about the bombers' target for as long as possible.

Hamilton duly took off and was not far from the target when he heard Cheshire on the radio, calling in the last of the other aircraft to drop their bombs. Hamilton told Cheshire, 'We're ten minutes away. Do you want me to bomb?'

'Yes, I'll put down another marker for you,' Cheshire said.

Hamilton's crew dropped their bomb, and though it was difficult to spot the impact for the dust and smoke because 'when these Tallboys go off, there is a terrific amount of smoke … we got very close.'[4]

Because the bombs penetrated so deep into the earth before detonating, none of the watching aircrews had seen more than pinpricks of light when their bombs exploded, but the eruptions of smoke, pulverised rock, chalk and earth that followed showed their devastating power. Three Tallboys hit the tracks and one penetrated the tunnel roof, drilling through 70 feet of earth and soft chalk rock into the tunnel itself, its detonation triggering a 10,000-ton rockfall that left a crater 25 metres wide at the surface and completely blocked the tunnel. The Panzer division was delayed for a crucial few days, allowing the Allied build-up of troops, tanks and equipment in Normandy to continue and the break-out from the beachheads to begin. So effectively had 617 Squadron done their job that the Saumur tunnel was not declared safe and reopened until three years after the end of the war.

Although he continued to fly daylight ops, the Saumur raid proved to be Nick Knilans' last night-combat flight. 'It came at a good time,' he said, 'because my subconscious mind was beginning to rebel at the continued stress.' Although he said he never had nightmares or even dreams about his wartime flying experiences, 'occasionally my hands would feel like they were filled with writhing maggots'. Daylight flying was much less stressful because it was easy to see the horizon and keep from becoming disoriented, whereas night-time evasive actions – violent dives and turns – left his senses reeling, and his 'dizzy feeling of falling out of control' made him struggle to believe the dials on his instrument panel.

Knilans was beginning to display the clear signs of overexposure to incredibly stressful situations, but what is today termed 'post traumatic stress disorder' (PTSD) was barely recognised in the Second World War. Known as 'shell-shock' in the First World War, it was now described as 'combat fatigue'. Psychiatrists believed that men in combat had a series of mental defences that were peeled away as the combat stress increased. The first defence was a belief in a distant ideal: 'the war to end all wars', 'fighting for democracy', etc. Nick Knilans had now lost his naive, youthful faith that the war would 'unite everyone in America and end all discrimination'. When talking to other American servicemen, he learned that African-Americans were allowed to use the swimming pools at their base only on Mondays. On Monday nights the pools were drained and refilled with fresh water before the white American servicemen used them. The implications were obvious. Knilans also wrote to his parents in Wisconsin to ask if the sign reading 'Gentiles Only' at the local country club was still in place. It was.

As a boy, Knilans had watched the local Ku Klux Klan, including one of his uncles, assembling around their burning cross in the park facing his house. 'They were mostly anti-Catholic,' he said, 'as there weren't any Negroes to harass. By the light of the burning cross, my brother, some friends and I would shoot stones at the Klansmen with our slingshots. My uncle was the tallest and the favourite target.' He complained to Knilans' father, but the boy was not punished and his father's only comment was to tell him that 'we should be sure and hit every Klansman at least once!'

With Knilans' belief in distant ideals eroded, the second layer of defences against combat fatigue identified by psychiatrists was to focus on a short-term objective, like the completion of a tour of duty. However, it was hard for any aircrew to focus on this when they knew that the odds of survival were stacked against them. One unpublished intelligence report gave a pilot a 10 per cent chance of completing two tours of combat flights, or, as Knilans observed, to put it another way, there was a 90 per cent chance of getting killed. 'I could keep my conscious mind from dwelling on the morbid thought. It must have been registering daily in my subconsciousness, though. So many men had been killed flying out of Woodhall Spa that I was losing track of them. I would go to a friend's room for a chat, only to recall on the way that he had been shot down.' Knilans felt that his self-confidence was 'weakening progressively ... I did not want to endanger my crew. I did want to get them safely through their second tour. Then they would be out of the combat part of the war.' Knilans did not know it at the time, but he was testing his luck, and his mind, to the extreme.

★   ★   ★

Ten days after D-Day, on 14 June 1944, 617 Squadron were tasked with leading a massive force of bombers in attacks on the hardened concrete U-boat pens at Le Havre, where German E-boats – high-speed surface craft capable of combat speeds in excess of 40 miles an hour – were sheltering. Armed with cannon and torpedoes, and operating under cover of darkness, the E-boats had already caused substantial losses to Allied shipping in the Channel, particularly the transport craft carrying munitions and supplies in support of the invasion.

Leonard Cheshire, in a Mosquito, dived down from 3,000 feet to 700 feet before releasing his red target markers. As he did so, he was 'completely encased in hundreds of rounds of light flak shells' but 'escaped one more time'. Twenty-two Lancasters of 617 Squadron followed him in, bombing with Tallboys, and by the end of the raid the pens and harbour were littered with wrecked E-boats, only one of them now fit to put to sea.

Despite the dangers, some aircrew revelled in the opportunity to see action even if they were not required to take part. Nick

Knilans' Lancaster was also hit by heavy flak, probably causing another 617 Squadron pilot, Jimmy Castagnola, to regret having come with him as a passenger, though in the event the only casualty was the mid-upper gunner, who was hit in the leg. Castagnola had volunteered to 'come along for the ride' even though it would not count towards his total of ops. He, and 'many other 617 types,' Knilans said, 'liked to go along just for the excitement of combat … These off-duty passengers liked the feeling of no responsibilities and the thrill of the danger encountered.'

The dangers were very real. The following night, the squadron was airborne again, this time targeting the E-boat pens at Boulogne. They had taken off in perfect weather and flew over the Channel through 'gin-clear skies', but as they approached the French coast they found cloud obscuring the target.[5] Squadron Leader Les Munro, leading the Lancaster formation, decided that the weather conditions made an accurate attack impossible and signalled a return to base.

In the event, Wing Commander Cheshire, flying a Mosquito, descended through the cloud and found the skies clear below 8,000 feet, the absolute minimum height for the Tallboys to achieve sufficient terminal velocity to penetrate the concrete structures. Cheshire overruled the order to return to base and the Lancasters turned back towards the target.

As Nick Knilans' Lancaster was completing his bomb-run, James Castagnola, again flying 'second dickey', 'just for fun', shouted, 'Look up, Nicky!' As Knilans did so, he saw 'a dozen thousand-pound bombs coming at us. Another squadron's Lancaster had dumped their bombs down through the clouds. I made a very steep diving turn to port. Roy, in the rear turret, claimed he could have patted one of the bombs as it hurtled past his turret.'

All the bombs missed and Knilans made it back to base unscathed, but 'Mac' Hamilton was lucky to make it back at all after being riddled by a barrage of flak as he began his bomb-run. 'The first shot hit us – the only reason I'm here to say this is because it went off under the Tallboy [which blocked most of the shrapnel], but it damaged the hydraulics.' The next burst damaged the bomb doors, though without hydraulics they couldn't have closed them anyway. The next one hit the starboard wing, and the starboard landing gear came down, causing such drag from the slipstream that it pulled the aircraft to the right. The next shot punched a tennis-ball-sized hole straight through the starboard middle tank and they lost 400 gallons of high-octane fuel, which sprayed out all over the rear gunner, the vapour almost incapacitating him.[6]

By this time the flak was getting even heavier and Hamilton started to turn off, but the bomb-aimer, Roly Duck, shouted, 'Hold it Mac, I want a picture!' so they carried on towards the target. 'It felt like a fortnight,' Hamilton said, 'but was only about fourteen seconds.' Just as he heard the bomb-aimer's exultant shout of 'We've hit it!' another flak burst tore the aircraft's nose off.[7]

The force of the blast badly wounded Roly Duck, and blew him back through his compartment and halfway up the steps to the cockpit. The whole of the glass nose of the cockpit had disappeared and the ferocious wind blew all the navigator's maps and papers and 'a load of Window all over the place'. The drag on the aircraft from the lowered landing gear and the hole in the nose was also causing them to lose height steadily. Crossing back over the Channel, they were down to about 3,000 feet and expecting to ditch somewhere between Boulogne and Folkestone.

The crew had no intercom and could communicate only by shouting or passing scribbled notes to each other. Two of them were trying to treat Roly Duck's wounds as he lay on the floor of the cockpit. Still in shock, the gaping wound to his leg felt to Duck 'like being hit on the back of the legs with a football on a cold day. My legs were numb.'

As his crewmate Flight Sergeant Leonard Rooke cut away Duck's trousers to examine the wound, Duck said, 'What the hell is wrong?'

'Well,' Rooke said, 'you've got a hole in the front of your leg and an even bigger bugger in the back.' He later confided to Mac Hamilton, 'I don't know what he's making such a fuss about. I could only find one hole, my finger only goes in that far,' pointing three inches from the end of his forefinger![8]

Duck felt a savage stab of pain as 'some bloody fool' – his crewmate Rooke – tried to inject him with morphine, and he kicked out. His flailing legs hit the throttle quadrant, causing both starboard engines to cut out and throwing the aircraft into an immediate steep dive. Knocking Duck's feet out of the way, Hamilton managed to regain control and restart the engines. 'Through gritted teeth, Duck told Rooke not to bother with the morphine.'[9]

As Hamilton flew on, out of the corner of his eye he glimpsed what at first he thought were fighters overtaking them 'at fantastic speed', even though one of them appeared to be on fire. Only later did he realise that he had just had his first sight of one of Nazi Germany's terror weapons: the V-1 flying bomb.

Desperately short of fuel, he then got a message from West Malling telling him to land there, but warning him that flares were being put down to guide him to a landing on the grass 100 yards north of the main runway, leaving it clear for the fighters based there which were chasing the flying bombs.

Hamilton's crew only managed to lower the port under-carriage by using the emergency air bottle to power it, and there was no way of knowing if it was locked down or not. They tried to use the rest of the air to lower the flaps, but it ran out when they had only been lowered halfway. Nonetheless, after ordering the crew to crash positions, Hamilton managed to make a safe landing on the grass. When he'd pulled to a halt, someone came to the front of the aircraft, looked up through the gaping hole in the nose and said, 'God, how did you get this thing back?'[10]

A medical team brought Roly Duck out of the aircraft on a stretcher. He was conscious but had lost a lot of blood, and his flying suit was soaked in it from chest to knees. The doctor who examined him took one look and cheerily said, 'Christ, I've never seen so much blood. Are you sure you haven't been hit in the balls, old man?' Surgeons later removed twenty-seven pieces of flak from him – a shortage of anaesthetic caused him to regain consciousness halfway through the operation – and presented them to him in a NAAFI mug as souvenirs.

The unfortunate Duck then developed gangrene in both legs. Doctors were planning to amputate, but then agreed for him to be transferred to the RAF hospital at Rauceby, near RAF Cranwell, in the hope that his legs might be saved by treatment there. Rauceby housed a specialist unit, where a group of surgeons including members of Archibald McIndoe's 'Guinea Pig Club' pioneered techniques to rebuild the shattered bodies and faces of crash and burn victims.

Mac Hamilton, thinking he would be doing his friend a favour, flew Roly Duck to Rauceby in a Lancaster and was focused on performing the softest of landings, determined not to cause his bomb-aimer any further discomfort. He was making a perfect landing when he felt a jolt and saw that the port main wheel had separated from the undercarriage and was rolling down the runway ahead of them. He fought to hold the aircraft upright as long as possible, but, inevitably, the wing dug into the ground, slewing the aircraft around, and fire broke out in both port engines.

The crew's exit from the burning aircraft was a model of speed and efficiency, and it was only when all of them were standing on the runway that they remembered the badly injured Roly Duck was still inside, strapped to his stretcher in the rear of the fuselage. Along with the shaken WAAF who had been escorting him, he was safely rescued from the aircraft and taken to the RAF hospital, where the leading orthopaedic surgeon, George Braithwaite, operated and saved both legs, albeit with the aid of bone grafts that left him with a permanent limp.

Reconnaissance photography after the attacks on Le Havre and Boulogne showed that, despite the difficulties with weather and the air defences, the raids had been outstanding successes. Several Tallboys had penetrated the concrete roofs of the pens at both sites, wreaking havoc among the E-boats berthed there. An estimated total of 130 E-boats had been destroyed, with an incalculable saving of Allied shipping and lives as a result.

CHAPTER 8

Terror Weapons

On 13 June 1944 the first pulse-jet-powered V-1 flying bombs had been launched against London. As the war progressed, these terror weapons became a massive threat, as Johnny Johnson recalls: 'I'd read the reports of the attacks on London. It was like a second Blitz and there was a sense of anger and a need to get back at the enemy.'[1] From the cockpit of his Lancaster, one 617 Squadron pilot had even watched two of the first V-weapons being launched. 'The flames from their base looked like a pale green telegraph pole,' he said, 'going straight up into the sky. Their flight path curved up, over and down into England.'[2]

For those on the ground, the experiences were horrific. One V-1 exploded near the BBC's Bush House on the Aldwych in central London during the busy lunch hour. 'It was as though a foggy November evening had materialised at the throw of a switch,' a BBC employee recalled. 'Through the dust and smoke, the casing of the bomb lay burning at the corner of Kingsway: three victims lay unmoving at the top of the steps, and figures were scattered all over the road.' In a nearby first-aid post, the supervisor saw the body of a young female colleague. 'She was naked and dead, stripped and killed by the blast. Another one I knew came in with blood spurting from her wrist and a deep gash in one eye. From 2.15 to 5.15 p.m. we were treating casualties.' The final death toll was forty-six, but another six hundred people had been injured in this single attack.[3]

Nick Knilans also saw the impact of the buzz-bombs at close quarters when he spent a spell of leave in London. He went alone – none of his crew would go with him because of the V-1 threat – but Knilans:

> *did not fear them enough to keep me away from the good times to be had there. If you were in a pub, you could hear the steady droning noise of one approaching. All talk would stop. Someone would say, 'Come on, you little bugger, come on!' You would be safe if the buzz-bomb flew overhead before its engine cut out. Then it would glide into a nearby area but, hopefully, not damage the pub that you were patronising.*

When he went to the cinema, a notice would regularly flash up on the screen, warning that the air-raid alert had been sounded, but he sat tight and saw 'less than twenty people out of two hundred' leaving to seek shelter. The all-clear notice would appear on the screen a few minutes later, but the whole procedure would be repeated several times during a single showing of a film.

Knilans had a very near miss while staying at the Regent Palace Hotel. He was going up in the lift to the sixth floor when it was 'severely jolted by some noiseless explosion in the hotel. The elevator dropped rapidly some forty feet before the emergency brakes stopped it. The operator and I said nothing. She just took me back up to my floor.'

As soon as the lift doors opened, he was enveloped in a thick cloud of dust and smoke. As he began to walk along the corridor towards his room he saw that every door had been blown off its hinges and the floor was carpeted with glass, debris and fragments of clothing. His own room was covered in broken glass from the shattered windows. The bomb had struck the floor above, killing or injuring several employees resting in their rooms.

The launch facilities needed to be destroyed at all costs. The next generation of more powerful and longer-range V-2 rockets was also being prepared, but the V-3 supergun, an even more fearsome weapon of mass destruction, was being readied at Mimoyecques, 15 kilometres south-west of Calais. Code-named *Hochdruckpumpe* (High-pressure Pump), the V-3 assembly and launch site was buried in a labyrinth of tunnels, galleries and chambers carved out of the chalk bedrock and covering an area of two and a half square kilometres, up to 100 metres below ground. Should the flak batteries clustered around the site not prove sufficient deterrent to air attack, the facility was also shielded by a massive concrete roof 6 metres thick, pierced by a series of narrow openings lined with 20 centimetres of armoured steel. In the space of nine months, 120,000 cubic metres of concrete had been poured on the site, building the protective roof and the network of galleries and stores for the weapon, its explosives and propellants and the 1,200 men who were to garrison the site.

The heavily camouflaged site was devoid of visible activity, above ground at least, but was being closely watched by Military Intelligence, who reported that German scientists were completing a devastating new weapon there, though as yet they did not know what the weapon actually was. On 6 July 1944, 617 Squadron was tasked with attacking the complex. The aircrews were called into an early morning briefing, showing that it was to be a daylight raid, and at the same time 100 Halifax bombers from Main Force were carpet-bombing Mimoyecques, dropping almost 500 2,000-pound bombs on the site. However, while they wrecked the above-ground railway line and pounded the surface buildings and flak batteries, the 2,000-pounders did minimal damage below ground level. To destroy the facilities deep below the surface would require 617 Squadron's ground-penetrating Tallboys.

Later that morning, Leonard Cheshire, now flying a single-seater American Mustang in preference to the slower British Mosquito, led a formation of the squadron's Lancasters armed with Tallboys to destroy the site, whatever it hid. Cheshire had first flown the Mustang operationally in a raid on a V-weapon site at Siracourt ten days earlier. Astonishingly, the first time he had ever flown one was earlier that same day. Don Cheney watched Cheshire:

> *teaching himself to fly this damn plane! He took off with the instruction book on his lap and the plane bouncing down the runway like a grasshopper.*

*There was an awful struggle to get the wheels up but he finally succeeded and
flew around for half an hour or so, then came back in and bounced a few
more times as he landed and taxied in. He kept us all in stitches over drinks
in the Mess that night, telling us about his struggles to find the bloody lever
to raise the wheels.*

Struggling with navigation over a blacked-out landscape, at
night and often in bad weather, Bomber Command crews often
got lost on their way to the target. Yet somehow Cheshire, sitting
in the unfamiliar Mustang for only the second time, was capable
not only of flying the aircraft at high speed and at low level, but
also of navigating himself so precisely across the featureless Channel and the flatlands of northern France that he arrived directly
over the target, smack on time.

The weather forecast for the Mimoyecques raid was clear skies
for take-off and over the target, and, unlike the long-range ops
targeting Munich or munitions factories in the south of France,
this was, said John Pryor, 'one of the raids that was over before
you really got into it. We had taken off, arrived, and were on our
journey home, all in about two and a half hours.'[4]

Although battered by Main Force, the camouflaged site was
still heavily defended by the remaining flak batteries, but, ignoring the flak, Cheshire dived down in his Mustang and laid his
red spot-fire markers directly onto the target from an altitude of
800 feet. As Cheshire circled, calling in the raiders one by
one, he saw one Tallboy score a direct hit, clipping the corner of
the concrete slab, while near-misses penetrated the earth around
it. In all, eight Tallboys struck the earth close to the target at
approaching the speed of sound, and drilled down deep below
the ground before detonating. The craters they created on the
surface were huge enough – between 25 and 35 metres in diameter and up to 15 metres deep – but most of the explosive power
was confined below ground, where it had a devastating impact,
caving in the vertical shafts and causing the subterranean tunnels
to collapse, destroying the weapon but, in horrifying 'collateral
damage', also burying alive the 300 slave labourers and their

guards working below ground. Those who were lucky enough to have been above ground never returned to the site, which was abandoned on 26 July. Another of Hitler's doomsday weapons had been destroyed.[5]

All the raiders returned safely, though pilot Ross Stanford barely made it back after losing two engines to anti-aircraft fire just after leaving the target. Stanford was an Australian cricketer who had made 416 not out as a fourteen-year-old but only ever played one first-class match. Run out without scoring while batting for New South Wales at the other end from Don Bradman, he was never picked again. Les Munro flew alongside him, escorting him all the way back to England. 'He was losing height the whole way,' Munro says, 'but we managed to make it back to a diversion airfield and landed together. I loaded his crew up and flew them all back to Woodhall.'

A week later, unaware both of the extent of the destruction that 617 Squadron had already wrought and of the German decision to abandon the site, and still fearing its reoccupation, USAAF commanders authorised a raid using pilotless radio-controlled Flying Fortresses, each loaded with 11 tonnes of TNT and Torpex High Explosive. As well as the V-3 site at Mimoyecques, they also targeted the V-1 and V-2 sites at Watten, Wizernes and Siracourt, but the op was a shambolic failure. On 12 August 1944 the US Navy attempted a further attack on Mimoyecques using a B-24 Liberator, but the aircraft exploded shortly after take-off, killing the crew, including pilot Lieutenant-General Joseph P. Kennedy Jr, eldest son of Senator Joe Kennedy, who had been grooming him for a run for the US presidency after the war. John F. Kennedy inherited the mantle and duly became US President in 1961.

On 5 September 1944, any potential German threat from Mimoyecques was finally eliminated when it was overrun and occupied by advancing Canadian troops. Even then, the nature of the site and the Nazi terror weapon it had housed were not conclusively established until after detailed inspections had been made by expert scientists and engineers in November and December of that year.

The experts concluded that what 617 Squadron had destroyed were fifty subterranean superguns – the largest guns ever seen. Two banks of twenty-five firing tubes, each 127 metres long and inclined upwards at an angle of 50 degrees, were sited 1,000 metres apart. At the deepest level, 100 metres below ground, galleries gave access to the clusters of breech-blocks allowing shells, fitted with steel fins to aid accuracy, to be loaded. Unlike conventional cannons, which were powered by a single explosive charge, each barrel of the supergun was fitted with a series of explosive boosters at intervals along the bore, like a series of interconnected, inverted Ys, that would detonate in sequence to increase the speed of the projectile. Accelerating all the way up the barrel, each shell would burst from it at a speed of 1,500 metres a second, striking London, 165 kilometres away, less than two minutes later.

Like the First World War 'Paris Gun', which was fired at the French capital during the spring of 1918, the V-3 supergun would have been too inaccurate to be effective against purely military targets and was specifically designed for terror attacks on civilian populations. It was capable of firing supersonic shells at a rate of almost 600 an hour, raining down almost 600 tons of high explosive a day on London. Although that daily total of high explosive was modest compared with the thousands of tons being dumped on Germany's cities by Britain's Main Force bombers and the USAAF, had the V-3 been used to attack London, the psychological impact on a population still recovering from the Blitz would have been devastating.

The Nazi forces were in headlong retreat on both Eastern and Western fronts by December 1944, and the threat that they might ever use the supergun had been eliminated, but Churchill still pressed for its complete destruction, fearing that a future enemy – a reborn Germany, the Soviet Union or even France – might one day reinstate it and use it against London.

The day after VE Day, 8 May 1945, which brought the war in Europe to an end, operating clandestinely to avoid alerting General de Gaulle's provisional French government, which had been resisting attempts to destroy the site, British Royal Engineers laid explosive charges inside the excavated and accessible parts of the Mimoyecques site. Five days later they laid further charges – a total of thirty-six tonnes of TNT in all – to complete the job, demolishing the upper levels and the concrete and steel carapace covering the site. The Second World War superguns had never been fired.

★   ★   ★

Led by 617 Squadron, the eighteen-month bombing campaign against the Nazi V-weapon sites and the transport infrastructure surrounding them had first reduced and then virtually eliminated the threat from Hitler's terror weapons. Sir Arthur Harris was able to claim that 'instead of an average of 6,000 flying bombs, the enemy was only able to launch an average of 95 a day ... judge what a bombardment more than sixty times as heavy would have been like!'[6]

One of the last acts of that campaign, the destruction of the Nazi V-3 supergun, proved to be Leonard Cheshire's final bow with 617 Squadron. He was compulsorily stood down from ops by Air Commodore Ralph Cochrane the next day. The normal tour of duty was thirty ops, and Cheshire had completed a hundred, many of which were the most perilous of all, at extreme low level. He also spent longer than anyone over the target, calling in each bomber in turn onto the markers he had laid. As testament to his courage, he was awarded the VC, the first ever to be awarded for a period of sustained bravery rather than an individual act.

At the same time as Cheshire was stood down, Cochrane also retired all three of 617's Flight Commanders: the Australian Dave Shannon, New Zealander Les 'Happy' Munro, and the American 'gentle giant' Joe McCarthy. Despite his nickname, Munro was definitely not happy about it. He had flown fifty-nine ops and was 'really disappointed to be stood down. I would have preferred the round figure of sixty ops. I never thought it would have been pressing my luck to go on, I don't think any of us did.

We were all very close-knit and ran a very efficient operation during the Cheshire era. He exuded confidence in his own abilities, in flying, in operations and in running the squadron, and that skill and professionalism filtered down to all of us.'

Cheshire and the three Flight Commanders were replaced by 'Three Limeys and a Welshman': Wing Commander James 'Willie' Tait took over as OC of the squadron, while Squadron Leaders Gerry Fawke, Tony Iveson and John Cockshott became Flight Commanders.

Tony Iveson grew up in York. His father had fought in the Great War and been wounded on the first day of the Somme. Like so many other veterans, he never talked about his experiences, but in the mid-1930s, when the talk of war was growing, he was 'in despair at the thought that it was all going to happen again'. Iveson had been 'crazy about flying' from the age of ten. He and a couple of friends used to save sixpence a week to buy *Popular Flying*, a magazine started by Captain W. E. Johns, author of the 'Biggles' books. Iveson's father told his son there was no future in flying and wanted him to 'find a proper career', but in 1938, aged nineteen, he joined the RAF Volunteer Reserve. 'I was tested the same day as the Munich crisis and I thought: There's no point in going home, best put on a uniform and get cracking. I wasn't naive in my expectations, I'd read all the old war books and knew about the horror of being shot down in flames; I'm not sure I thought of honour and defending my country, but I did feel that, unlike previous "political" wars, this one really had to be fought.'[7]

By 1940 he was flying Spitfires, but in May of that year he fell ill with appendicitis and 'was late' getting into the Battle of Britain. 'Most of the men I'd been training with were killed,' he said:

> and I probably would have been too but for my appendicitis. We were so short of pilots in 1940 that there was no time for proper training – you just had to get in and do it. You did some formation flying and aerobatics, you did one or two height climbs to get used to being on oxygen, and that was

about all. There were no dual-control Spitfires, so when you got on a squad-
ron, if it was in Action Theatre, you were almost straight in the air.

Still raw, he joined 616 Squadron in late August 1940 and on
16 September, on patrol over a convoy in the North Sea, he chased
down a German Junkers Ju 88. 'I had been told you had to get in
close, but foolishly I got too close,' he recalled. The rear gunner
poured fire into Iveson's Spitfire and he had to ditch in the sea. 'I
don't think we were ever instructed about ditching, but I remembered vaguely from chat in the crew room that you landed along
the waves and not into them.' As he hit the sea, the Spitfire 'bounced
about and I was thrown around, then she hit again and came to a
stop. I stepped out and the plane disappeared.' He only had a Mae
West life-jacket, no dinghy, but he was spotted and picked up by a
minesweeper with the convoy. 'I was incredibly lucky to get out
alive,' he said. 'I don't know many others who did after ditching
in a Spitfire. That was my first experience of combat, five days
after my twenty-first birthday, and it was a real eye-opener.'

When he got back on dry land, he was taken to Coltishall,
where he was simply given an aircraft, which was needed at his
base, to fly home:

That was the welcome back, no medical check-up or anything – just get
on with it. I didn't have a parachute or a helmet, but I flew the Spitfire
back to my squadron. It was all very casual, but so was the atmosphere on
an operational squadron in those days. Young pilots would arrive and
probably no one even knew their names, and often they didn't come back
from their first or second trip.

After completing his tour with 616 Squadron, Iveson spent
two years as a flying instructor in Rhodesia. Still only twenty-one
when he arrived there, he mentored eighteen-year-olds who had
never even sat in an aeroplane before, and within four weeks he
had them flying solo, by both day and night. However, he wanted
to play a bigger part in the war effort, and in August 1943 he
returned to England.

He was determined to go back to Fighter Command and protested loudly when he was posted to Bomber Command instead, but his new CO sat him down and told him, 'Look here, young man, the war has changed. Fighter Command did its stuff in 1940, but now the only Command which is taking the war to the enemy is Bomber Command. We want good pilots, you're a good pilot and that's why you have been selected.' Iveson recalled:

> I hope this doesn't sound conceited, but although I was new to bombers, flying was second nature to me by then. It must have been very different for a young pilot going to a Main Force squadron and probably taking over a bunk which somebody had disappeared from the night before. A lot of those young guys were still trying to learn to fly – to cope with weather and icing, and the enemy of course – and their first trip might have been Berlin.

After Iveson completed his training at the Lancaster Finishing School, the chief flying instructor asked him if he would like to apply to 617 Squadron. He didn't need to be asked twice:

> 617 was something special. It had a reputation as the premier squadron, though some said it was a suicide squadron because of the losses on the dams. It was a big thing to be chosen. I couldn't believe it, I was very proud. The day I arrived at Scampton my Flight Commander looked me up and down and said, 'Christ, a sprog like you, never done a bomb op in your life; they'll have you for breakfast.'

They nearly did, one day later. 'There was no messing about on 617,' Iveson said:

> Not like Main Force, where they had to take care of COs and Flight Commanders. If there was an operation on 617, everyone went. We had the advantage over the Army of not seeing people being shot, mutilated, blown to bits alongside us, coughing their guts out. So someone being missing was not too dissimilar to someone being posted. You'd come back from leave and find that two crews had been posted or you'd get back from

a trip to find your friend had 'bought it'. We didn't talk about it, we'd just say, 'Poor old Joe got the chop' or 'Another one's gone for a burton.' We certainly didn't do what they did in the film Battle of Britain *– put a wreath in front of the Mess table – there was none of that sentimental nonsense. It was accepted as part of life at the time. To go on a series of operations and not expect anyone to get into any kind of trouble would have been stupid.*

On ops, Iveson tried to keep his focus solely on the task in hand:

On the bombing run, I would always put the seat down and concentrate on the instruments and let other people look out. I was watching my six instruments and that little extra one we had for the SABS bombsight which gave us our direction. I was trying to ignore what was going on outside and get on with the job. I felt that I had a responsibility to myself and the squadron to do as good a job as possible, but as the skipper, I also had a responsibility for the crew: to get them there and back safely. You expected flak over a target and it was much preferable to fighters. You knew you had to fly through it and there was nothing you could do – you couldn't take avoiding action if you were going to do a good bombing run – so it was pointless worrying about it.

When off duty they used to go out together as a crew, and Iveson remembered an army colonel being 'very surprised' when his sergeant used his first name to ask, 'Would you like another pint, Tony?' Iveson said to the colonel, 'We're a crew. We fly together, we know each other and we depend on each other. On the station and in the air it's different – more formal – but here we're off duty.'

While Iveson and the other new commanders were settling in, Nick Knilans and his flight engineer, Ken Ryall, were flight-testing aircraft. Some were new, others had been patched up after taking battle damage. The two men usually flew without any other crew, though they did not entirely lack for company since, as Knilans said, 'several times we took along a friendly WAAF. [Flying Officer] Phil [Ingelby] and I took turns flying and wooing.

After all, the RAF had supplied the Lancaster with a couch!' He was referring to the rest bed fitted in some Lancasters.

★   ★   ★

Amidst the frivolities of RAF life, 617 Squadron continued to target the V-weapon sites, including the enormous concrete block-house and tunnel at Watten, where V-2s were to be assembled and stored. The world's first ICBMs, V-2s were liquid-fuelled and, travelling at four times the speed of sound, would climb to a height of 75 miles before plunging down onto London and the south-east at supersonic speed. Even New York was thought to be within the V-2's potential range. Unlike the V-1 'doodlebugs', there was no audible warning of a V-2 strike at all, as they flashed down from the skies much faster than sound could travel. The first, launched on 8 September 1944, killed three people in Chiswick: an elderly lady, a three-year-old girl and a sapper on leave from the Royal Engineers. Between then and 27 March 1945, when the last V-2 struck London, they were to kill 2,751 people in the capital alone.

Don Cheney, who was on leave, had a ringside seat as one of the first landed near the train on which he was travelling. 'There was a terrific explosion,' he says, 'and you could feel the shock-wave pulse over the area. I didn't see it, but I could certainly feel it!'

Barbara McNally lived in a pub in Camberwell that suffered a direct hit. They did not use the air-raid shelter in their street, preferring to use the pub cellar, where they pushed their beds together so that the two children, their aunt and their parents could all sleep in relative safety. Or so they thought.

The bomb blew out the side walls and the building collapsed onto us. Luckily one of the girders in the roof of the cellar created a small space above us to stop us all being killed. My dad was the only casualty, as the till from the bar above fell on his head and he was very bloody and unconscious. I remember my mum and aunty both crying and praying and shouting for help. I thought I would help by pushing all this debris that was in front of me out of the way. That nearly brought the house down, literally. I don't know how long we were down there but eventually the emergency services heard our shouts and we were pulled out more or less feet-first and put in the shelter in the street.[8]

One of the most horrific V-2 attacks came on 25 November 1944 when a rocket hit a crowded Woolworth's store in southeast London. Witnesses described the store bulging outwards and then imploding in a blinding flash of light and an enormous roar. People several hundred yards away felt the heat of the blast on their faces. The Co-Op store next door also collapsed, killing more customers inside. The bodies of passers-by were flung great distances, and an army lorry overturned, killing its occupants. A double-decker bus was spun round, causing more death and injury; the passengers could be seen still sitting in their seats, covered in dust.

Only piles of masonry and body-parts remained where Woolworth's had once stood. It was to take three days to clear the debris and retrieve all the bodies. In the carnage 168 people had been killed and 121 seriously injured.[9]

★　　★　　★

The Watten V-2 site also housed a huge reinforced-concrete factory for producing liquid oxygen, almost five tons of which were needed to launch each rocket. Although the Watten site, protected by a 16-foot concrete roof, was not completely destroyed, repeated attacks on the site itself by 617 Squadron using Tallboys, and on the surrounding road and rail tracks by USAAF bombers with conventional bombs, disrupted production and threatened a catastrophic explosion of the liquid oxygen compressors and tanks, forcing the Germans to abandon the site for another V-2 assembly and launch bunker in a disused quarry outside the village of Wizernes, near St-Omer.

Set into the quarry-face, the site was capable of handling rockets of up to twice the V-2's 50-foot length. Air reconnaissance had first detected construction at Wizernes as far back as August 1943, and by January 1944 an elaborate system of camouflage had been installed on the hilltop in an attempt to conceal the site and its underground facilities. The Wizernes site was protected by a huge concrete dome, a cupola 16 feet thick, 230 feet in diameter, and weighing 5,500 tons. A bomb-proof ferroconcrete 'skirt' supported by a series of buttresses extended beyond the dome, giving added protection, and dispersal tunnels protected by enormous blast-proof doors led to a series of concealed launching sites, each about the size of a tennis court, scattered through the surrounding countryside.

In March 1944 Wizernes had been added to the list of targets for Operation Crossbow – the bombing campaign targeting all V-weapon sites – and over the following three months the USAAF and RAF carried out a series of raids, dropping 4,000 tons of bombs without causing any significant damage to the complex, though the constant air-raid warnings did stop construction over 200 times in May 1944 alone.

The Germans believed that no bomb could destroy the concrete dome, but Barnes Wallis had other ideas. He conceded that hitting the dome directly would have required an almost impossible degree of accuracy, and even Tallboys could simply bounce off the domed surface. However, Wallis had already argued persuasively that a near-miss with a Tallboy could be just as damaging, and possibly more so, than a direct hit, and if there were near-misses at Wizernes, the subterranean earthquake effect of the bombs would be enough to collapse the dome and block or destroy the underground tunnels, galleries and chambers.

Such was the secrecy involved in planning the attacks – coupled with the lack of knowledge about what the sites actually were – that most of the crews had no real idea that they were targeting V-weapon sites. On their first attack at Watten, the target was described as a 'power station', and 'certainly no one told us it was a V-2 site,' John Bell says. 'I don't remember any target ever being described as a V-2 site, it was just another target. We'd head in towards them and just attack the relevant markers.'

617 had been given the task of destroying the Wizernes complex using Wallis's ground-penetrating Tallboys dropped from 17,500 feet to obtain maximum terminal velocity. Bombing from that height called for a very high degree of skill. Even with the SABS to assist accuracy, it was 'like putting a bomb in a barrel'.[10] Three previous raids had been unsuccessful, largely because of poor visibility over the target. On one daylight raid on 20 June there was so much dust and haze in the air that the results of the attack were inconclusive, though as they were returning to base, pilot John Pryor came up with the idea of the 'gaggle' system. It was a fairly compact formation but with each bomber at a different height and relative position to its neighbours, making them a more difficult target for flak batteries, giving all the aircrews the freedom to bomb without interference from those around them and also allowing each aircraft to make a direct run over the target without being buffeted by the slipstream from the one ahead of it. The gaggle formation was used on all subsequent 617 ops.[11]

The squadron returned to Wizernes for a night raid on 22 June, but thick cloud obscured the target completely and they were ordered back to base without dropping their bombs. On 24 June they tried again. Aerial reconnaissance had reported light cloud over the target area earlier in the day, so take-off was delayed until late afternoon.

Gerry Hobbs was a twenty-one-year-old from Guildford who had been a messenger boy in the Auxiliary Fire Service in Islington before joining up 'because everyone else was – the war was on and it seemed the right thing to do'. He was the wireless operator in a Lancaster piloted by Flight Lieutenant John Edward as, in a cloudless sky, they climbed over the Channel. Heading towards the target, they could still see the English coastline behind them. They were on their straight and level bombing run when two flak bursts hit the aircraft, setting both port engines on fire. The flight engineer was killed instantly and, but for the huge bomb beneath them that absorbed much of the shrapnel, the flak might have claimed more victims.

Gerry Hobbs

'When we were hit by flak it was so sudden,' Hobbs says:

I could feel it peppering the side of the aircraft. A piece ripped past me tearing a piece from my parachute pack; you don't think about it like this at the time, but I was only inches from death. That's how fine the line was. You were quite isolated at your own station in the Lancaster – you didn't really see what was happening in the rest of the aircraft – so all I knew initially was that the wing was on fire and we were in trouble.

Hobbs quickly switched from his radios to the intercom, and after the navigator and pilot had made unsuccessful attempts to extinguish the fires, he heard the pilot's call of 'Abracadabra! Jump! Jump!' – the prearranged signal to abandon the aircraft.

We used that so there could never be any misunderstanding, because we'd heard of other crews baling out by mistake and the pilot bringing a half-empty aircraft back to base, but there was no doubt here – we were going down. There was no time to put out any calls on the radio, time simply

*ceased to exist in the normal way. It all became automatic. I was surprised
at how calm I was, to be honest.*

Hobbs could not later recall he and his crewmates ever speaking about the possibility of being shot down or killed. 'Privately, I suppose there was always the thought of getting the chop at the back of your mind – there were a hundred and one ways: accidents in bad weather, mid-air collisions, flak, fighters, being hit by someone else's bombs – but our main focus was on getting the job done and getting home to the pub!'

Now, for the first time, he had to face the reality. There was 'a split-second panic' as he tried to decide whether to take off his flying helmet, but then he moved aft, keeping his arm across his torn parachute pack from which silk was already billowing in the wind roaring through the holes in the fuselage. As he climbed over the main spar to reach the exit door, he met the Canadian mid-upper gunner, J. I. Johnston, coming the other way. Johnston shouted that the rear of the aircraft was ablaze and there was no escape that way. They hurried forward back over the main spar, feeling the aircraft shuddering as they did so. The flight engineer was lying on the floor, 'clearly dead, but there was no time for regrets; we had to get out of this aircraft or we were all going down with it.'

As Hobbs stepped over the flight engineer's body, he saw the pilot, Flight Lieutenant John Edward, leaving his seat. 'He didn't see us and I didn't distract him, but I remember releasing his oxygen tube.' As Edward made for the front hatch, the last thing Hobbs saw was the pilot checking his harness, and then Hobbs passed out from lack of oxygen.

When he regained consciousness, he was lying in a cornfield surrounded by German soldiers and French civilians. A column of smoke was rising from the burning wreckage of the aircraft in the next field. 'The first thing I remember is coming to and seeing faces staring down at me,' Hobbs recalls. 'There had always been a chance that I'd come face to face with the enemy. Now the time had come. I wondered how they would react.'

He looked at the Germans and said 'Deutsch?' They said 'Ja' and that was 'the limit of our conversation. Things were hazy. I must have banged my head on landing, as I had a broken nose as well as a broken right arm and leg. And the parachute must have dragged me, as my face, legs and hands were covered in cuts and grazes.' He pointed to his top pocket and a Frenchman took out his cigarettes and lit one for him. Hobbs's arm and leg were splinted and he was laid on a bed of straw on the floor in the back of a lorry and taken to hospital in St-Omer. Beside him lay the bomb-aimer, John Brook, and the Canadian navigator, Lorne Pritchard, who was alive, but 'wrapped in his parachute with a lot of blood on it'.[12] All three men became PoWs.

The Frenchman who had lit Hobbs's cigarette was André Schamp, a member of the French Resistance, who had helped many British soldiers trapped in France after Dunkirk to escape through the Pat O'Leary organisation. After the war he was personally awarded the Croix de Guerre by General de Gaulle. He had been working in a garden in the nearby village when he saw the bomber coming down in flames. He ran across the fields towards it but was then 'terrified, as the aircraft was right above me and whichever way I ran, it seemed to follow me'. When no more than 150 metres above him, it exploded in mid-air and crashed 200 metres away. 'After the explosions, some parachutes came down like candles.'

In spite of the danger from exploding ammunition, he began to search for survivors, finding first the pilot and another crewman, who were both dead, and then the unconscious mid-upper gunner, Johnston. Schamp ran home to get a first-aid kit, taking Johnston's revolver with him, which he hid. By the time he got back, Johnston had also died. The Germans had now arrived, but though they were suspicious that Johnston's holster was empty when there was ammunition in the pouch, Schamp escaped further investigation that would surely have led to his death.

He began searching again and found Hobbs sitting in a field of oats, with his head barely visible above the crop. Schamp lit a cigarette for Hobbs, who thanked him in English and then said in French 'Vive la France.' There were no other survivors, for the rear gunner was dead, still trapped in his smouldering turret, and the other crewmen could not be found. The wounded men were taken to hospital by truck and became prisoners of war. The Tallboy bomb had not exploded and lay 'flat in the ground to a depth of two metres'. It too was loaded onto a truck and taken away for examination by the Germans.

Schamp had to wait until late that night before the Germans allowed him to leave, but the next morning, as he came out of a church service in the village, a German army truck pulled up and he was ordered to remove the bodies of two of the dead airmen, the pilot, Flight Lieutenant Edward, and the rear gunner, Flight Sergeant Samuel Isherwood.

Schamp and his wife laid them on a cloth on the church floor and organised funerals for the following day. They tried to bribe a German guard with snacks and drinks to allow them to also bury the other gunner still trapped in his turret, but it took three visits to the German regional HQ to get the necessary permission. The dead gunner was 'in a terrible condition and difficult to remove from his turret', Schamp said, 'but my courageous wife laid him out as best she could'. They dug three graves in the churchyard and made oak coffins for them. Because of the bombing, the electricity had been cut off and the coffins had to be made by hand. They did not finish them until 1 a.m. The next morning 'a beautiful service was read in my own Roman Catholic religion' and the bodies were buried. A month later the body of another crew member, Flying Officer W. J. King, was found about 200 metres from the crash site, apparently flung there by the force of the impact. He was buried at Longuenesse Military Cemetery.

Hobbs didn't discover the fate of his pilot until after he returned home at the end of the war. 'I always carried the hope he had survived. But soon after I got home I discovered he'd been killed. I don't actually remember anything, but I presume that he must have pushed me out of the hatch but then couldn't

get out himself,' Hobbs says, his voice faltering. 'That's something that has been on my mind for the last seventy years.'

Almost forty years later, accompanied by pilot John Edward's
sister and cousin, Gerry Hobbs met André Schamp for the first
time since that brief encounter. 'Their hospitality was overwhelming,' Hobbs says. They visited the V-2 site at Wizernes
they had been trying to bomb when they were shot down, before
making a sad pilgrimage to lay flowers on the graves of Hobbs's
dead comrades.

★   ★   ★

Although a number of near-misses were recorded during the raid,
causing damage to the railway lines and buildings around the site,
and one pilot claimed to have seen a bomb penetrating the dome,
Wizernes had again escaped fatal damage, and the squadron was
sent back to finish the job on 17 July 1944, the first op under the
leadership of Wing Commander James 'Willie' Tait, who had
taken over command of 617 Squadron five days earlier.

Tait didn't like the nickname 'Willie' at all, but it had stuck to
such an extent that a request to meet 'James Tait' would have
been met with blank looks from most of his squadron. Ruddy-
faced and still only twenty-six, Tait was not a gregarious character like his predecessor. His perfect evening, according to some
of his men, would have been sitting in a chair with his eyes
closed, listening to classical music, not socialising in the Mess.
He could appear aloof, even cold, but that was mainly a product
of his shyness. Larry Curtis felt that he 'could never get as close
to Willie [as Cheshire]. I don't think anyone really did so …
maybe because he was a very shy person … but when he got in
the air he was very much a pro.'[13] However, others found him a
more approachable character. 'Almost every evening, unless
there was something brewing,' Arthur Ward, his wireless operator, said, 'he'd be in the bar with a pint tucked under his arm and
an old pipe going.'[14]

Whatever his social graces, in the place that really mattered – in
the air – there was general agreement that he led from the front

and was a worthy successor to the great Leonard Cheshire. Tait was, said one aircrewman, 'a great CO, courteous, not frightening. He welcomed me to the squadron, but after that you didn't really have any dealings with him on day-to-day squadron life. He was always leading from the front, an impressive man who we readily followed and looked up to.'[15] His rear gunner added that Tait was 'a dream to fly with, and this is especially noticeable in the rear turret, the best place to judge a pilot's skill, believe me, where his feather-touch three-pointers hardly made the turret shudder.'

Tait had been obsessed with speed and fast cars since childhood – when he grew up, he owned a succession of sports cars and broke his leg when the starting handle of one kicked back as he was cranking it – but when his father took him to see the Schneider Trophy Air Race in 1928 he decided on the spot to become a pilot.

When he joined 617, he had already flown almost a hundred ops, plus 'many others not recorded', and had a DSO and Bar and a DFC to his name. By the time he left the squadron, he had added another two Bars to his DSO and a Bar to his DFC. His reputation as a brave and brilliant pilot extended far beyond 617 Squadron. 'He had so many DSOs and DFCs,' Australian pilot Bruce Buckham of 463 Squadron said, 'that it's a wonder to me that they didn't give him a VC to go with them.'

Tait, flying the Mustang Cheshire had used, and Gerry Fawke in a Mosquito were to mark the target at the V-2 site at Wizernes for a force of sixteen Lancasters carrying Tallboys, while another Mosquito would film and photograph the aftermath of the bombing. It was a short flight to the Pas de Calais on a beautiful day, and 'the view was amazing,' Nicky Knilans recalls. 'We'd had a couple of aborted trips because of cloud and now it was totally clear.' Just fifty-nine minutes after take-off, Tait swooped down to 500 feet to mark the target through 'a heavy hail of light flak and machine-gun bullets'.[16] Fawke added two more markers ninety seconds later. Flying with 500-foot height separations, the Lancasters had all been 'flying in a circle until the markers went in, then we all turned in together'.

As John Bell squinted down through the bombsight, he could see that the surrounding area was pockmarked with craters from previous attacks by 617 and Main Force. Bell could see the quarry, vehicles and railway lines and then 'a small dot … this dome at the end. On the ground, it is massive, but of course from 18,000 feet it was just the tiniest of pimples on the earth.' The target was visible at all only because the grey concrete dome stood out clearly against the dusty white of the surrounding chalk. Bob Knights made a good level run in, and Bell felt the aircraft lift as he released the bomb. 'I watched it fall all the way,' he says, 'and of course, as it neared the ground, it still had a lot of forward motion so it actually looked like it was "flying" across the countryside, racing along.' As he saw it explode right beside the dome, he shouted 'Bullseye!', elated with the achievement of seeing it strike right on target. When they got back on the ground at Woodhall Spa, Bob Knights' crew were all 'still on a high', says Bell. 'It was full circle: all that training, all that preparation, a great crew, great aircraft, an accurate bomb and bombsight, all coming together perfectly at the right moment.'

Knights' crew usually celebrated another successful op by going to Boston or one of the local pubs. However, John Bell was now in a relationship with a WAAF called Florence who worked in the Intelligence Section, and so 'tended to do my own thing a bit more'. He was at a dance in the Sergeants' Mess, 'quite shy and standing around, not quite knowing what to do', when he was introduced to Florence, and they hit it off straight away and started spending a lot of time together.

The squadron intelligence officer became used to Bell popping in to the Intelligence Section and pretending to look at maps so he could see Florence. 'There was an intensity to it all,' Bell says. 'It provided a real diversion from the enormity of what we were all doing on ops and the hectic, intense aircrew lives we were leading: living together, flying together, drinking together, and sometimes dying together. Florence and I would just get on our bikes, go off to the pub on our own and talk about normal things.

Never for a meal, mind, we couldn't afford that!' Before long they were engaged.

★   ★   ★

Several of the other bombs dropped on the V-2 site had also detonated virtually simultaneously, multiplying the earthquake effect. 'All we saw was this huge mushroom cloud over what had been the target.'[17] The near-misses penetrated the ground and blew out the supports for the concrete dome, which collapsed, along with a huge section of the old quarry face. Barnes Wallis's theory that a near-miss could be more damaging than a direct hit had been proved. German reports stated that 'the whole area around has been so churned up that it is unapproachable and the bunker is jeopardised from underneath', and the officer in charge of the site reported to his superiors that 'Persistent air attack with heavy and super-heavy bombs so battered the rock all around that in the spring of 1944, landslides made further work impossible.'[18]

The Wizernes site had been put beyond repair, and was abandoned without a single V-2 rocket ever being launched from it. However, the destruction was deep underground, and at surface level the site looked little altered, which led 5 Group commander Ralph Cochrane to order another attack three days later, on 20 July. This time Tait marked the target with two smoke markers, but cloud and ground haze were drifting over the target and Nick Knilans told Tait that from 18,000 feet he could not distinguish the markers' twin plumes of smoke from the general cloud and haze. In an action that was supremely brave, or suicidally foolhardy, while bullets and flak fragments continued to bombard his aircraft, Tait then began to fly his Mustang in a tight circle around the target, calling 'Bomb on me!' over the radio.

Fearing that his bomb might hit Tait's aircraft, Knilans told his bomb-aimer, who could 'just barely' see the Mustang, 'I don't like this, don't bomb.' He radioed to Tait, 'We can't see you. Unable to bomb.' So did the rest of the squadron. 'Maybe they were as appalled as I was at the act of bombing Willie and his Mustang,' Knilans later said. 'No one ever talked about it. We just turned around and took our Tallboys home again.'[19]

The Wizernes V-2 site

The collapsed dome at Wizernes, like a giant concrete mushroom, still stands, and in 2013 John Bell returned to the site with the author almost seventy years after the raid. At the age of ninety, he braved the blazing sun to climb the steep, bramble-strewn slope of the dome to see exactly where his 12,000-pound Tallboy had struck. 'I watched this point through my bombsight seventy years ago,' he said. 'Now I'm standing on that same spot. What an experience! I could never have imagined, back then, that seventy years later I'd stand in the same place! It really is one of the best moments of my life.'

CHAPTER 9

Mac's Gone!

25 July 1944 dawned a bright and sunny summer day, and 617 Squadron returned to the V-2 site at Watten. The German flak was intense, and Don Cheney's Lancaster took a succession of hits – 'you could hear it pinging off the side of the aircraft' – and was badly damaged. The aircraft was awash with leaking hydraulic fluid and a cloud of blue, acrid-smelling cordite smoke from the flak shells that were exploding so close to them that they could hear the 'Boom!' and feel the plane shudder from the blast. One engine was knocked out and the hydraulic system was so severely damaged that the gun turrets were inoperative, the bomb doors couldn't be closed and the landing gear couldn't be lowered.

The intercom was working, fortunately, but, when checking in with each member of the crew, Cheney was unable to obtain a response from the mid-upper gunner, 'Mac' McRostie, and sent the wireless operator, Reg Pool, back to investigate. Pool was back within a few seconds, looking very shaken. 'Mac's gone!' he shouted in Cheney's ear.

'Gone where?' Cheney said, puzzled.[1]

'He's baled out! I got to the rear door just as his flying boots disappeared outside!'

Cheney put his aircraft into a gentle banking turn away from the target, and sure enough, 3,000 or 4,000 thousand feet below them, clearly outlined against the green fields, he saw a parachute gliding gently down. There was nothing to be done, so Cheney kept turning and set course for home. Part of the Perspex above the cockpit had been blown out and there were numerous holes in the wings and fuselage, but there was no fire and none of the remaining crew had been wounded.

As they reached the French coast more flak came up. 'We took such evasive action as we could,' Cheney says, 'with so much of our "laundry" hanging out, but fortunately the flak bursts drifted past harmlessly and we began a steady descent in order to increase airspeed and get out of enemy territory as soon as possible.' As he was doing so, Flight Engineer Jim Rosher tapped him on the shoulder and pointed upwards.

There, not more than 15 feet above them and sliding gently to port, was 'the most beautiful Spitfire I have ever seen!' Cheney says, smiling at the memory. 'The Spit slid back and forth above and below us for some time until we were well over the English Channel. Then he perched off the starboard wingtip for about five minutes, grinning and giving us the thumbs-up; then, with a saluting gesture, he peeled off to starboard and was gone.'

Cheney considered landing at one of several airfields near the coast, but decided they could make it back to their own 'roost', and calling for 'special consideration' from the tower, he was cleared to come straight in. As they made the final stages of their approach, they managed to blow down and lock the landing gear using the emergency air bottle, avoiding the need for a more dangerous 'crash landing'. They touched down and coasted to a stop. They were escorted to the nearest dispersal pad by a retinue of fire trucks and ambulances and, still fearing a fire or explosion, as soon as they came to a halt Cheney ordered the crew to evacuate the aircraft.

Once safely on the ground and away from the aircraft, Cheney turned to look at his Lancaster. They had been flying P–Peter – not their regular aircraft but 'borrowed' from a comrade who was on leave while their own was undergoing a service check – and the flak storm they had flown through had left it riddled with nearly a thousand holes. 'The last I saw of it', Cheney says, 'was when it was parked behind one of the service hangars, where mechanics were busy salvaging as much of it as possible. A new machine was sent in to replace it, and, on his return, my colleague reluctantly accepted my apologies for doing away with his beloved P-Peter! It also cost me a good few at the Mess bar in order to assuage his crew!'

Although he didn't realise it at the time, Cheney himself had also been hit by flak. A couple of days later, he found a large lump on his right shin just below his knee. He then remembered that in the heat of battle over the target, his right leg had suddenly been knocked off the rudder pedal and there had been a burning sensation like a bee sting. It had faded after a few moments and he had then forgotten all about it. There was a scab on his leg and his battledress had a trace of dried blood at the same spot, but seeing nothing else amiss, he promptly dismissed it. However, a few days later, the swelling was getting bigger, the wound was sore and itchy and turning purple, so he went to see the medical officer.

The MO took a quick look, swabbed the wound with alcohol and asked his WAAF nursing assistant to hold a white metal tray close to his hand. With a pair of pincers he then extracted a jagged piece of black flak-shell fragment which had lodged itself against the shin bone and let it drop with a clang into the tray. Having bandaged the wound, he rummaged about, found a small cardboard box, lined it with cotton wool, laid the shrapnel fragment in it as if it were a precious jewel and put another layer of cotton wool on top. He then handed it to Cheney and slapped him on the back.

Cheney kept 'my very own piece of a deadly German 88mm flak shell as a keepsake of a very lucky escape! It reminded me of that very narrow line between life and death.' It was a very narrow line, and Cheney would be reminded just how thin that line could be as his tour of ops at the cutting edge of Bomber Command progressed.

★ ★ ★

During much of July 1944, the Allied advance from its Normandy beachheads had been painfully slow as they fought 'The Battle of the Hedgerows' with retreating German forces. It was named for the terrain that made Normandy such forbidding territory for invading troops. The small fields, surrounded by thick, high hedges and flanked by sunken lanes and ditches, were a nightmare for Allied tanks to cross, while providing perfect cover for enemy machine guns and anti-tank weapons. Only on 19 July were Allied troops able to begin the break-out from the hedgerows into more open country where the advance could accelerate.

Before D-Day, most of 617 Squadron's ops had been at night, but in the aftermath, with the Allies enjoying progressively greater air superiority, more and more raids were made in daylight. On 31 July, 617 Squadron was briefed for a daylight attack on a storage tunnel housing V-weapons at Rilly-la-Montagne. Once more they would be supported by Lancasters of 9 Squadron. 617 were to bomb the tunnels at the southern end, while 9 Squadron targeted the northern end. They were all armed with Tallboys fitted with time-delay fuses, allowing the bombs time to penetrate the tunnel roof before exploding, increasing the damage caused and making salvage operations more difficult. Once the Tallboys had been dropped, 300 Lancasters from the Main Force, each carrying twelve 1,000-pound bombs with delay fuses, were to carpet-bomb the tunnel from end to end, aiming to ensure its complete devastation.

One of the 617 Squadron crews was piloted by Flight Lieutenant William Reid, VC, the son of a blacksmith from Ballieston near Glasgow. Even with the moustache he cultivated, he seemed far younger than his twenty-two years, though the look in his eyes showed he had seen and done things that would have destroyed a weaker man. Reid had been awarded the Victoria Cross while with his previous outfit, 61 Squadron, for his heroism on a raid on Düsseldorf in November 1943. Soon after crossing the Dutch coast, his Lancaster was attacked by a Messerschmitt. His windscreen was shattered and he suffered wounds to his head, shoulder and hands from shrapnel and jagged shards of shattered Perspex. The aircraft's communications system and compasses were put out of action in the attack and the elevator controls damaged, making the aircraft difficult to control. Although the rear gun turret was also badly damaged, the gunners managed to drive off the attacker and, saying nothing about his own wounds, Reid checked that his crew were unscathed and then flew on.

Soon afterwards, the Lancaster was again attacked, this time by a Focke-Wulf Fw 190. It was a very unequal contest. The Fw 190 was almost twice as fast as the Lancaster, with a top speed of over 400 miles an hour, a climb rate of 2,500 feet a minute – well over three times the best the Lancaster could achieve – and its 30mm cannon were 'powerful enough to destroy most heavy bombers with just two or three hits'.[2] The German pilot raked the length of the bomber with his cannon, killing the navigator, fatally injuring the wireless operator, wounding Reid yet again and wrecking the mid-upper turret, leaving the aircraft defence-less. The oxygen system was also destroyed but, even though he was also hit in the forearm, the flight engineer kept Reid supplied with oxygen from a portable cylinder.

Reid refused to turn back and, having memorised his course to the target, reached Düsseldorf – one of the most heavily defended targets in Germany – fifty minutes later. The failure of the com-munications system meant that the bomb-aimer knew nothing of Reid's wounds, nor the casualties among his other comrades. The Lancaster's camera showed that when the bomb-load was dropped, the aircraft was directly over the centre of the target.

Reid then turned for home. With no navigator, he plotted a course using the Pole Star and the moon. Already badly wounded, frozen by the wind roaring through his shattered windscreen, weak from blood loss and half-blinded by the blood from his head wound running into his eyes, Reid now had to fly without oxygen, as the emergency supply had given out and, dodging heavy anti-aircraft fire over the Dutch coast, he did not want to descend below the 10,000-foot ceiling for a flight without oxygen until he was safely clear of the flak batteries that might have finished the job begun by the fighters.

When he eventually descended to lower altitude, the warmer air caused his facial wounds to reopen, half-blinding him again with trickling blood. He kept lapsing into semi-consciousness, but the flight engineer and the bomb-aimer helped him to stay awake and the Lancaster airborne as Reid nursed his crippled aircraft back across the North Sea. He made an emergency landing at the Ship-dam USAAF base in Norfolk, despite ground mist that partially obscured the runway lights. Even though half the undercarriage collapsed as he touched down, Reid brought the aircraft to a halt without it overturning or catching fire and without further injury to himself or the surviving members of his crew.

His Victoria Cross citation stated:

> *Wounded in two attacks, without oxygen, suffering severely from cold, his navigator dead, his wireless operator fatally wounded, his aircraft crippled and defenceless, Flight Lieutenant Reid showed superb courage and leadership in penetrating a further 200 miles into enemy territory to attack one of the most strongly defended targets in Germany, every additional mile increasing the hazards of the long and perilous journey home. His tenacity and devotion to duty were beyond praise.*[3]

After recuperating from his wounds, Reid at once volunteered to join 617 Squadron in January 1944. On his first flight with 617, he made a hash of a landing and damaged his tailplane. Leonard Cheshire told him that he blamed himself for not giving him a few circuits to get him back in the groove after the trauma of his last op before sending him off solo, but then added that he'd have to put a red endorsement in Reid's logbook. As Reid himself noted with a rueful grin, 'I think I'm the only pilot to get a Victoria Cross on one trip and a red endorsement on the next!'

With his comrades, he would now be bombing the V-weapon tunnel at Rilly-la-Montagne from 12,000 feet, ensuring a clear sight of the target, whereas the Main Force would drop from 18,000 feet. When told that at the briefing for the op, Reid immediately expressed his concern, saying: 'In that case we will be flying through their aiming point and the line of the drops.' However, he was assured that the Main Force would not begin making their bombing runs until twenty minutes after 617 had completed theirs. The 617 Squadron aircrews were also told that their Mosquitos would be covering the necessary photography of the target, so 'if the flak is bad, don't bother with pictures'.⁴

It was a clear, bright, sunny day and the flight to the target was incident-free. As Nick Knilans began his bomb-run, his flight engineer tapped him on the shoulder and said, 'Look up above you, Skipper.' Knilans' irritation at having his bomb-run interrupted disappeared in an instant as he saw another Lancaster 250 feet above him with its bomb doors open and about to drop its own load. Knowing his own aircraft would be invisible to the crew of the one above him, Knilans swerved out of its path using full left rudder then settled on a new course for the target quickly enough for his bomb-aimer to find the aiming-point and release their Tallboy. As Knilans swung away to begin the run back to base, thanking his stars for the narrow escape from being bombed by one of his own comrades, his rear gunner shouted, 'Skipper, one of our planes just had a bomb dropped on it.'

From its altitude of 16,000 feet, the doomed Lancaster took four minutes to fall to earth. Knilans asked his flight engineer if he could see any parachutes. 'Not yet,' he said. 'Oh! It's breaking up! There! I see one – no two chutes. There it goes into the ground.'

Knilans' crew all fell silent, knowing that while two men had escaped their doomed aircraft, the other five were now dead. That aircraft had been piloted by Bill Reid, VC. As he completed his bomb-run, Reid had seen flak ahead and was 'of a mind to turn off immediately', but his bomb-aimer, Les Relton, called, 'Hold it.' The next instant, Reid felt the aircraft lurch violently. There was none of the acrid smoke or smell of gunpowder that came when hit by flak, so he knew that, as he had feared, they could only have been struck by bombs falling from above, as someone in the Main Force began bombing well ahead of their scheduled time.

The bombs – 1,000-pounders – smashed through the Lancaster's port wing and central fuselage, tearing out an engine, severing the control cables and fatally weakening the aircraft. When the control column went 'loose and floppy' Reid ordered, 'Stand by to bale out!' His engineer at once passed Reid his parachute pack before strapping on his own. As the aircraft's nose dipped, sliding into a dive, Reid knew he had no means of controlling it and shouted, 'Bale out!'

The flight engineer, Sergeant James Norris, rushed forward, trying to get out of the escape hatch, but the aircraft had gone into a steep dive and began to spin, making it difficult for Reid to move. His one thought was to jump as soon as possible, but at first he struggled to get out of his seat because the control column was now jammed against the chute pack on his chest. He fought with the sliding windows on either side of the cabin, but neither would open. The aircraft was still spinning down in a near-vertical dive. Reid finally struggled out of his seat but couldn't get to the forward escape hatch. Then he remembered the dinghy escape hatch above and a little behind his head. He dragged himself towards it, every step a huge effort as the aircraft dived almost perpendicularly towards the ground.

Just as he took a grip on the handle of the hatch and began to turn it, he heard banging, rattling sounds and suddenly found himself 'in freefall', with the only noise the whistle of the wind. The nose and cabin of the aircraft had simply disappeared, torn free from the main fuselage, and Reid had tumbled out of the hole.

He felt for his ripcord, pulled it, and his chute opened with a savage jerk. He saw woodland beneath him, coming up quite fast, and 'kept my legs together, as all good Scotsmen are told to do!' He hit the canopy of a tree, slid down the main branch and hit the trunk. The impact left him with a badly broken wrist and severe bruising to his left leg, but he was otherwise unhurt and managed to work his way down the trunk to the ground.

He hid his Mae West in the undergrowth, put a shell dressing on his broken and bleeding hand, and then began to move south, away from the sound of the delayed-action bombs that were continuing to detonate. However, he had not travelled far when he was captured by a group of German soldiers. As they were taking him to their headquarters, about a mile away, he saw the wreckage of part of his aircraft and asked them to let him look at it. The bodies of the mid-upper and rear gunners were lying on the ground alongside it. His flight engineer, Norris, had also been captured, but Reid harboured hopes that the remaining three crew members had made good their escape. 'I was certain I was the only one trapped in the plane when spinning down. It was only that the nose of the plane came off or I'd never have lived myself.'[5] Reid and Norris became PoWs, and it was only when they returned to the UK at the end of the war that they discovered that all five of their comrades had died, unable to escape from the aircraft as it plummeted to earth.[6]

Another 617 crewman saw 'several unfortunate incidents where a stick of bombs from one of our aircraft hit another, which peeled over and then collided with another Lancaster. It was terrible, there were bodies tumbling through the sky. Nowadays it's called "friendly fire" – we had a lot of that on daylight raids.'[7]

Colin Cole

Wireless Operator Colin Cole's Lancaster was another victim of 'friendly bombing':

A strange article smashed through the window of the Lanc and hit the wireless equipment. It fractured the hydraulic oil pipe and I got smothered in oil. There was smoke and a terrible burning smell – it was an incendiary bomb dropped by another aircraft above us! We were just so lucky it didn't ignite or that would have been the end of us all. Imagine that, surviving the German onslaught and then dying at the hands of your own mates! I was in hospital for about a week after that because I lost the sight in my eyes, but as soon as I was fit, I was sent up flying again.

Rilly-la-Montagne proved to be 617 Squadron's final attack on a V-weapon site, because as Allied troops pushed ever further eastwards, they eventually overran them. Examination of the captured sites by intelligence officers revealed just how devastating 617's precision bombing with Tallboys had been. Despite their massive reinforced-concrete roofs, with the assembly and launch sites buried deep underground, the V-weapon sites at Watten, Wizernes, Creil, Rilly-la-Montagne, Siracourt and Mimoyecques had all been devastated, their underground chambers, tunnels and galleries collapsed by the earth tremors generated by the exploding Tallboys.

CHAPTER 10

Life and Death on 617 Squadron

While 617 Squadron were working to eradicate the V-weapon threat, a plot to eradicate the greatest threat of all was narrowly failing. On 20 July 1944 Colonel Claus von Stauffenberg placed a bomb hidden in his briefcase under the table of the briefing room at the Wolfsschanze (Wolf's Lair), Hitler's military headquarters on the Eastern Front. Von Stauffenberg then left the room, but when the bomb detonated, although four men in the room died and most of the others were wounded, Hitler was shielded from the blast by the thick leg of the solid oak table, and escaped unscathed. The conspirators were rounded up and von Stauffenberg was executed by firing squad the same night. Eight other conspirators, including his elder brother, suffered a much more agonising fate after being repeatedly half-strangled with a wire garrotte and then resuscitated before they were allowed to die. The grisly events were filmed for Hitler's entertainment.

Elsewhere Allied advances continued, with Soviet forces liberating Majdanek concentration camp on 24 July and Florence liberated on 4 August 1944. A few days earlier, on 26 July, a German Me 262 fighter became the first jet-engined aircraft to shoot down an enemy when an RAF Mosquito – previously one of the fastest aircraft in the skies, but now a helpless victim of the super-fast turbojet – was hit by the Messerschmitt's guns and forced to crash-land.

With the V-weapon threat to London largely eliminated, 617's next series of targets were mainly naval ones, principally the havens of the U- and E-boats still threatening the Allied ships supplying the invasion forces in France. However, so many

Tallboys had been used against the V-sites that stocks had been exhausted. British factories were only producing seven Tallboys a week, and though American factories were turning out four times as many, shipping delays and losses to U-boats led to shortages. As a result, the raid by 617 Squadron on a bridge at Etaples on 4 August saw each Lancaster carrying twelve 1,000-pound bombs, and though the area surrounding the bridge was peppered with bomb-craters and several direct hits were recorded, the bridge still stood.

Back at the Petwood after the raid, the crews helped themselves to tea and biscuits and fought for one of the two copies of the day's *Daily Mirror* – highly prized for its pin-up picture on the centre pages and the cartoon adventures of the even more scantily clad Jane. The aircrews whiled away the time until dinner smoking and chatting, playing cards, darts or billiards or strolling in the grounds, then congregated in the bar for a pre-dinner drink. A couple of days earlier, Canadian pilot Don Cheney had enjoyed a 'rousing crew party' in the Mess for three newly commissioned Pilot Officers in which their brand-new officers' dress hats had been filled with beer and then stamped on repeatedly to signify that the officers were now 'fully operational'.

But on 4 August 1944, with an op scheduled for the following morning, there were no such raucous scenes for Don and his crew, nor any prospect of a run out to the Red Lion at Stickford or one of the other local pubs. The aircrew ate their dinner of 'steak' pie – actually horse meat – sprouts, mashed potatoes and gravy, and the inevitable steamed pudding with custard, and then, after a couple more beers, most of them were in bed before ten o'clock, knowing that they would be woken at four-thirty the next morning.

It was a gentle awakening, with a young WAAF bringing them tea and biscuits. They washed and shaved and put on their battledress, making sure to empty their pockets of all items, even letters, ticket stubs or banknotes, that might prove useful to the enemy if they were shot down. However, one aircrew member used to slip a small German dictionary in his pocket – 'Come in

handy in the Stalag' was his only comment.¹ They ate the traditional pre-op breakfast of bacon and fried eggs with fruit juice,
toast and coffee, and then boarded the crew buses for the short
journey to the airfield.

The briefing that morning of 5 August 1944 revealed the target
for the day. Sufficient Tallboys, fitted with special hand-tempered
noses, had now been delivered for a daylight attack to be launched
on the formidable hardened U-boat pens at Brest. In addition to
the usual flak and ground fire, and the fear of fighter attacks, there
was an added hazard in attacking Brest: the naval guns sited around
the harbour, which sent up heavy shells that exploded with a deafening 'Boom!'

The intelligence officer told them that squadrons of USAAF
Lockheed Lightnings and RAF Spitfires would be providing air
cover and that latest reports indicated that there were about seventy
88mm anti-aircraft guns around the U-boat pens, only a handful of
which were radar-guided. Even by the standards of Military Intelligence, this turned out to be a very optimistic assessment. French
Resistance sources claimed the defences included 175 88mm guns,
almost all of which were radar-guided. With the warnings of heavy
opposition fresh in their ears, the crews headed to the aircraft.

'The worst part of any op was sitting on the grass by your
aircraft waiting for the green Very light to set you going,' pilot
Lawrence 'Benny' Goodman remembers:

> That was the time I'd often think about what was to come, what we were
> about to do and about to face. There was nothing else to occupy your mind,
> so perhaps you might dwell on your immediate future, but the crew certainly
> never spoke about it or expressed any personal fears. As the captain, I would
> have stopped them if they had. I'm sure we all must have been apprehensive
> at times, but you couldn't let that feeling pervade the crew.

All Londoner Goodman had ever wanted to do was be a pilot.
'I didn't have any real notion of fighting,' he says, 'what I'd do
or be involved in, I just wanted to fly.' He was afraid he'd be
rejected as a pilot because his eyesight was less than 20/20, so

when he went for his eye test he took the precaution of learning the bottom line of the chart on his way in. He reeled it off and passed the test.

Benny Goodman and crew

Goodman volunteered for the RAF just after the start of the war in 1939, but it was over a year before he was called up.

I was only eighteen, but when I said I was going to join up, my father quietly encouraged me. He had served as a soldier in the First World War on the Eastern Front where the Turks and Bulgarians were fighting. He'd served from the start to the end, and been wounded in action, but he didn't really talk about the war other than to say it was tough fighting; very few prisoners were taken out there. When I told my mother I was joining up – Oh heavens! I was an only child and she did not want me to go at all, probably like all mothers at the time.

I'd seen the results of the German bombing – one night a bomb hit the buildings next door to our block of flats. It blew the place to smithereens and we had to evacuate down the fire escape. I still remember slipping on a piece of flesh at the bottom of the steps; I've never forgotten that.

I joined up because I thought it was the right thing to do in the circumstances. Everyone else was joining, so I didn't want to be the only one not 'doing my bit'. I just thought it was going to be a great adventure ... boy,

The Möhne dam: one of the iconic images from the aftermath
of the raid that earned 617 Squadron its nickname.

Two survivors of the Dams raid: Les Munro (far left) and Johnny Johnson, along
with Mary Stopes-Roe (the daughter of Barnes Wallis) and Wing Commander
David Arthurton, in front of a Lancaster bomber at a ceremony commemorating the
seventieth anniversary of the Dams raid.

Johnny Johnson in March 2014, and some seventy years earlier at RAF Scampton with his Dams raid crewmates. Front row, left to right: bomb-aimer Johnny Johnson; navigator Don MacLean; pilot Joe McCarthy; wireless operator Len Eaton. Back row, left to right: front gunner Ron Batson; flight engineer Bill Radcliffe.

Frank Tilley in February 2014. Tilley's first night op on 617 Squadron, a raid on the Dortmund–Ems Canal, was a baptism of fire: 'I was transfixed by it,' he says, 'and by seeing other Lancasters going down in flames … In that moment I realised what the war, Bomber Command, the flying, all meant.' Making a safe return to RAF Woodhall Spa, where 617 Squadron was based from January 1944 until the end of the war, he had never been so glad to see the runway.

The raid on the Gnome-Rhône aero-engine factory at Limoges in
Occupied France in February 1944 was the first op for John Bell and the rest
of Bob Knights' crew after joining 617 Squadron. Leonard Cheshire and
Mick Martin carried out the target marking, seen here.

John Bell and the author at the site of the V-2 assembly and launch
bunker at Wizernes in July 2013, standing just a few feet from where
Bell dropped his bomb on the reinforced concrete dome.

Colin Cole in March 2014. When 617 went back to the *Tirpitz* on
12 November 1944, wireless operator Cole watched his crew's Tallboy in flight
through the open bomb doors: 'It was an amazing sight to see it dropping
away, tracking to the target.' He would have seen something like the
photograph taken from Bob Knights' Lancaster on the same raid.

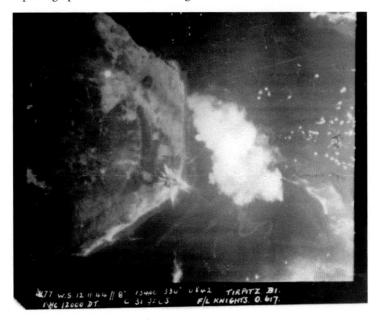

Willie Tait's crew after returning from the sinking of the *Tirpitz*.

The author standing in front of the wreckage of Bill Carey's 'Easy Elsie', which, having made seven runs at the *Tirpitz*, was fatally damaged by flak. Carey nursed his Lancaster over the Swedish border before crash-landing deep in the Arctic Circle. Although Carey was believed killed, and his kit cleared away by the Committee of Adjustment, he and his crew miraculously survived.

Iris and Sydney Grimes in March 2014, some seventy-five years after they met.

The decision to dispense with wireless operators and mid-upper gunners in order to accommodate Barnes Wallis's Grand Slam meant that Grimes's last op was the raid on the Bielefeld viaduct.

John Bell back in the bomb-aimer's compartment of a Lancaster, 'Just Jane', at the Lincolnshire Aviation Heritage Centre in July 2013. The same day brought together the author (centre) and, left to right, Benny Goodman, David Fellowes (460 Squadron), Colin Cole and John Bell, in front of 'Just Jane'.

did I get that wrong! I got into a bit of trouble during training, wacky fly-pasts and a few other antics, so I just burst out laughing when they told me I was posted to the famous 617 Squadron. I stopped laughing when I realised they were serious.

But my first time under fire was a real eye-opener. I could actually smell the flak! My wireless op was hit in his boot by a piece of shrapnel, and my bomb-aimer had a piece of cockpit above his head blasted away as a chunk came through it, missing him by inches. We also had an engine shot up, but boy, oh boy, could those Merlins take a battering. They gave such a beautiful roar when you pumped them up; there was no other sound like it in the world. They saved my life on more than one occasion.

Goodman would rely heavily on his aircraft, and his skills, as the days wore on.

The aircrews were to start their engines at nine o'clock, with all aircraft due to be airborne by half past. As they rumbled down the taxiway and queued at the end of the runway ready for take-off, the gunners were making last-minute checks, swinging their turrets from side to side and raising and lowering their gun barrels; if there were problems, it was as well to know of them now rather than when a German fighter had them in its sights.

Don Cheney's aircraft was fourth in line for take-off. His crew had extended their tour past thirty ops and were now up to an official total of thirty-nine. 'I actually counted it as forty,' he says, 'but the RAF didn't count Operation Taxable on D-Day as an op, as we hadn't been over enemy territory!' Unlike several other crews, Cheney's crew practised regular bale-out drills while safely on the ground, but he never imagined it would happen for real. It was rare for any aircrew to talk about being shot down, even though they knew that the possibility was always there and what the odds were.

Few aircrew had rehearsed escape and evasion techniques either, even though, back in July, Leonard Cheshire had decided that all his aircrews should practise escaping from the enemy by making their way back to base after being dumped in the country-side some miles away. Cheshire ordered them to assemble wearing their battledress and flying boots. He told them that the police had been instructed to arrest any airmen they came across, and then, after confiscating their caps and any money they had, he put them all on a bus with blacked-out windows. After a long ride involving many twists, turns and changes of direction to disorient them, groups of two or three men were dropped off at intervals.

Les Munro and David Wilson decided that their drop-off point was roughly due east of Woodhall Spa and took the first westbound road they came across. Nightfall and the descent of a typical Lincolnshire 'pea-souper' didn't help their navigation, but they ploughed on, circling through the fields to avoid a village where a patrolling policeman was obviously on the lookout for stray aircrew. Soaked to the skin, exhausted, and with blistered feet from the route march in flying boots, they eventually stumbled up the drive to the Petwood to find the place in darkness. Munro remembers feeling 'a bit peeved' that there was not even a welcoming committee to congratulate them for having made it.[2]

Others found a rather easier route home. Nick Knilans, his navigator, Harry Geller, and his wireless operator, Les Knell, were deposited in a narrow lane next to a small canal in a flat, featureless stretch of countryside, devoid of houses or buildings of any sort. As the bus disappeared from sight, Knilans turned to Geller and said, 'Well, Harry, where the hell are we?'

'Beats me, Skipper,' he said, 'without my Gee box and maps, I'm as lost as you are.' They followed the canal in what they hoped was the right direction until they came to a promising-looking road, and were wandering along it when they saw a car approaching. Fearing it was the police, they were about to sprint off across the fields when the smiling driver, one of Knilans' former comrades on 9 Squadron, stuck his head out of the window and said, 'What are you up to, Nicky?'

He offered them a lift to Boston and they all jumped in. 'Just drop us off at the White Horse Inn,' Knilans said. Having established with the barmaid that their credit was good, they remained in the pub till closing time and then borrowed the money for their bus fare back to the base. They were a mile or so from safety when uniformed local police flagged them down. Luckily, because of the blackout there were no lights on the bus, and by the time the policemen had boarded it and shone their flashlights towards the back, the three airmen had jumped out of the back door and disappeared into the night. They walked the last mile back to the Petwood and had gone to bed long before Les Munro and the rest of their comrades, including Don Cheney, came straggling in, having had a far less enjoyable escape-and-evasion exercise.

When Don Cheney had first started on ops, 'thinking I was a bit of a tough guy, I used to take a long cheese knife tucked in my flying boot as some sort of protection if I ever ended up shot down and on the ground, facing the enemy, and needed to defend myself or aid my escape. But after a couple of trips, I decided it might be better not to be carrying a knife if I ended up surrounded by armed Germans!' Cheney wasn't particularly superstitious and never took any lucky charms with him, as some pilots did – Mick Martin always carried a small toy koala bear – but 5 August 1944 would have been a good day to start, because Cheney was going to need a lot of luck to survive.

As the third aircraft began accelerating down the runway that morning, Cheney was already swinging his Lancaster, V-Victor, onto the apron of the runway, stopping at the white hash marks at the start point. As the previous Lancaster lumbered into the sky and banked around to port, Cheney eased the throttles forward, 'holding the straining V-Victor against the brakes with my toes pressing hard on the tops of the rudder pedals'. He released the brakes and, with the throttles against the stops, the Lancaster hurtled down the runway, the nose wheel glued to the white centre-line as the flight engineer called out the rising airspeed. Cheney could feel the aircraft straining to lift off despite its massive load, but waited until the airspeed reached 130 knots before easing back on the yoke. The Lancaster rose smoothly from the tarmac, and Cheney banked to port and began the long climb to the bombing altitude of 17,500 feet.

They crossed the south coast near Beachy Head, altered course to clear Cherbourg and swung in over the north coast of Brittany, where they rendezvoused with the USAAF Lockheed Lightnings. It was a clear, sunny day as the raiders flew on across the Brittany peninsula on a course that left the Germans guessing until the last minute about which of three potential targets they were making for. As they reached their final turning point for the attack, there had been no enemy activity at all – no fighters nor even any flak. It was, thought Cheney, going to be 'a piece of cake. We would lay our Tallboys right on the button.'

They switched course abruptly, heading for Brest, leaving the defenders little time to ignite their smokescreen. However, they were now flying at a precisely assigned altitude and speed, straight and level for 18 miles. The Germans knew the Lancasters' altitude and course and exactly where they were heading and put up a wall of flak so dense that, says Cheney, 'I'd never seen anything like it.'

The smoke markers dropped by Willie Tait from 4,500 feet fell into the water of the harbour and were 'rendered useless', but nonetheless, in near-perfect visibility, the bomb-aimers of the twelve aircraft that reached Brest acquired the target visually, and half a dozen of them obtained direct hits from three and a half miles above the target, though they had to fly through a blizzard of flak to do so.

Normally, flak pinged off the side of the aircraft, rattling like stones against a corrugated-iron shed. This time was very different. Don Cheney was in the final stages of his bomb-run, just seconds from releasing the bomb and listening to the calm, clear voice of his bomb-aimer Len Curtis talking him in to the aiming point, when the flight engineer nudged him in the ribs and gestured at the sky ahead of them. Still with his eyes glued to the instrument panel, Cheney could see in his peripheral vision an intense barrage of flak exploding in front of them and knew that they were just going to have to fly through the centre of it. 'It was radar-predicted, so the gunners knew exactly which piece of sky we were heading for,' he says, 'but as we were on the bomb-run, there was nothing to do but hope and pray as we headed towards it. I said, "Hang on guys, it looks a little dirty ahead."'

Puffs of grey smoke from the flak bursts drifted past on either side, the shell bursts so close that he could hear the muffled boom and glimpse their fiery red centres. Suddenly 'there was a real clang as we were hit,' he says, and then another and another – five or six in all. He heard cries of pain over the intercom and knew it was bad. The clang was from flak hitting the bomb – 'How it didn't go off still amazes me.' The aircraft filled with clouds of blue cordite smoke as it shuddered under the impact of two more direct hits. Yet more flak then erupted around them and there was another cry over the intercom, followed by Curtis's calmer voice saying, 'Bomb gone!'

Cheney at once banked the aircraft to port, pushing forward on the column to dive and increase speed, the fastest way to escape the danger, but the flak damage already looked fatal. Both gunners, bomb-aimer Len Curtis and flight engineer Jim Rosher were unhurt, but wireless operator Reg Pool was seriously wounded, slumped in his seat, eyes wide and skin very pale. Navigator Roy Welch was also badly wounded and unable to speak, though he managed to scrawl the course heading that Cheney should follow on a scrap of paper and held it out to him.

Cheney continued the turn away from the target and out towards the sea, but the starboard wing was riddled with flak holes – 'the biggest hole, a man could have climbed right through,' he says – and flames and dense black smoke were billowing from it. As he applied the left rudder, causing the aircraft to side-slip and blow the fire away from the fuselage, the starboard outer engine failed and flames belched from under its cowling. Jim Rosher feathered the propeller and activated the fire extinguisher, but the blaze in the number-two fuel tank could not be extinguished, and even if the tank did not explode, the flames were still burning along the wing and would soon engulf the fuselage.

'I knew almost immediately the aircraft was doomed,' Cheney says. 'It was all over and we were going down. The most important thing to do now was to get everyone out. There was no panic, everyone knew what they had to do.' He gave the order to abandon the aircraft and 'can still hear the voice of rear gunner Noel Wait calling over the intercom: "Wait for me! Wait for me!"' Cheney reassured him, but was still 'staring in horror at the flaming wing and gritting my teeth in expectation of the explosion of the fuel tanks. I had seen several crew survive such an event, but the prospect was terrible to contemplate.'

Cheney held the aircraft as steady as he could while the rest of the crew baled out. 'I felt huge responsibility to my crew and couldn't bear the thought of leaving anyone behind. They were my very good friends, men I'd flown with, fought alongside, socialised with, lived with – my family – and it was a shocking sight, to see some of them with terrible wounds.'

Jim Rosher passed Cheney his parachute and then helped Len Curtis with the forward escape hatch, which had jammed, partially blocking it – it was only 22 inches wide when fully open. They managed to force it open wide enough to squeeze through. Blood was running down the navigator's face as he stood in the aisle by Cheney's seat, waiting for his moment to go down through the escape hatch. 'He looked at me and smiled,' Cheney says. 'I smiled back and then he was gone. I still see that smile today. I was so proud of my crew.' Curtis and Rosher had waited to help the wounded navigator through before escaping themselves, while the two gunners escaped through the rear hatch.

Cheney and the badly wounded wireless operator Reg Pool were now alone in the aircraft. Cheney held the control column with one hand while reaching back to help Pool to his feet. The wireless operator shook his head at first but then gradually crept forward. He slumped against the side of Cheney's seat as the plane began to wallow from side to side and the nose dipped, increasing the speed. Cheney had to wrestle the aircraft back under control, then rose from his seat to clip on Pool's parachute. Each time he let go of the control column, the aircraft began to go into a dive and he had to switch rapidly between the controls and his injured crewman. He pointed to the ripcord and asked Pool if he could pull it and received a weak answering nod. Pool wriggled down into the hatch and signalled to Cheney to save himself. 'I saluted him,' recalls Don, 'then turned my attention to my own fate.'

The heat in the cockpit was now intense: 'I could see yellowish-brown bubbles and blisters breaking out on the flight engineer's panel. The heat was increasing and I was certain that the aircraft would blow up at any moment.' He grabbed the controls as the aircraft again began to dive, and wrestled it back into level flight. However, he knew there was no possibility of stabilising it long enough for him to clip on his parachute, make his way to the hatch in the nose and squeeze through before the aircraft went into a death dive so steep that he would be unable to escape at all. His only hope was to use the dinghy escape hatch in the roof above and behind him. He buckled his bulky para-chute pack on to his chest, knelt on his seat and wrenched the handles of the hatch cover open.

As it flew off into the slipstream, the already deafening roar of the wind blasting through the shattered aircraft, and the thunder of the remaining engines, grew even louder. He tore off his sun-glasses and flying helmet and got first one foot and then the other on to the seat, so he was standing in a crouch, with his knees bent. Putting one foot at a time on the armrests of the seat and straight-ening up, he was able to get his head and shoulders into the hatch, but the parachute pack was too bulky for him to get through while wearing it. He crouched down again, pulled the control column towards him with his foot to bring the aircraft back out of the dive into which it was slipping, then worked the chute pack out of the hatch first and got his head and shoulders through. Hauling with his arms and pushing with his feet, he managed to wriggle up until he had one knee on the outside edge of the hatch.

The aircraft was again slipping into a dive, and with the control column now out of reach, he had only seconds to escape. He groped with his foot for the back of the pilot's seat, braced him-self against it and then pushed with all his might.

As he shot out of the hatch, the slipstream caught him and hurled him towards the tail of the aircraft. Fortunately the mid-upper gunner had depressed his guns before abandoning his post, or Cheney would have been impaled on them. He missed them and the sharp-edged tail fins as he tumbled in the slipstream, his only injury a cut to his cheek as he skimmed the aerial wires from the fins. The smoke, flames and the deafening noise faded.

All that was left was the whoosh of the slipstream and the sensation of tumbling as he fell through the air.

'I remember seeing blue sky and white clouds, then green and brown farmland, then the sky again above my flying boots. I counted to three, grasped the ripcord and pulled with all my might!' There was a sickening pause, and when Cheney looked at the ring and two feet of wire in his hand, 'I thought, Good Lord! I've pulled too hard and broken the chute!' Mercifully, the canopy snapped open, he was jerked upright and found himself drifting down in the sunshine in complete silence.

'I was alone in the sky,' Cheney says, recalling his incredible escape from the comfort of his home in Canada, 'and wondered if I was dreaming; could this all *really* be true?' The burning Lancaster was still visible half a mile away, trailing smoke and flames. It had somehow come out of its dive and was climbing at a very steep angle, but as he watched, the nose suddenly dropped again and with its remaining engines still thundering, it spun down in a near-vertical spin and smashed into the sea. Flames, smoke, steam and spray were hurled high into the air and debris was spattered over a wide area, but as the sea subsided again there was only silence. The aircraft had disappeared, leaving nothing but a drift of black smoke that rapidly dispersed on the breeze.

There was no sign of any other parachutes, but it had taken Cheney so long to escape from the aircraft that his crewmates must already have splashed down in the sea. Raising his gaze, he saw the rest of the squadron's aircraft heading home. 'My comrades, disappearing over the horizon, would soon be safe back on friendly soil, but here I was heading towards enemy territory. I wondered if they'd raise a beer to us later that night.' One of his friends later told him that he'd watched through his binoculars from the astrodome of his own Lancaster, unable to tear his gaze away, as Cheney's aircraft was shot down. He had counted six parachutes leaving the aircraft, but had not seen a seventh, and with a heavy heart had resigned himself to the thought that Cheney had been unable to escape before his aircraft had crashed in flames.

Cheney splashed down into the sea and released his parachute at once, fearing it might otherwise drag him under. His Mae West inflated, and as he floated in the water he looked around, taking his bearings. He was about 18 miles south of Brest, just off the beaches of what, in peacetime, had been a popular French resort. He was a good swimmer, but hampered by his Mae West he could make only slow progress towards the shore. As he did so bullets began to strike the waves around him as German soldiers on the beach took potshots at him. He changed direction and began swimming back out to sea.

There was bright sunshine but a stiff breeze, and he had been in the water for about an hour and a half when he saw an old wooden fishing boat chugging through the waves towards him. 'Looking at this fishing boat approaching, I presumed I was about to be captured and that my war was over,' he says. 'It really was a contrast: the water was beautifully warm and serene, the French coast looked sun-drenched and beautiful, but I was about to fall into enemy hands. I was lucky to have got that far, but now it was all over.'

A few moments later strong arms were pulling him into the boat. None of the six-man crew was wearing German uniform, and when, in his best high school French, he asked, '*Où sont les Boches?*' one of them replied, '*Ah ha, les Boches kaput!*', dragging his finger across his throat for added emphasis.

That triumphant note was a little premature, because although American troops were advancing on the area, the German garrison remained in control of the fishing port where Cheney was being taken. Soon after the boat reached the jetty, he was hurried into a battered old truck and, flanked by heavily armed Resistance fighters, driven to a house a mile or so away and hustled up a flight of stairs. He was handed over to a Frenchwoman, who gave him some of her husband's clothes, fed him a sandwich and some fruit and then left him to sleep in a top-floor room.

Five hours later, as night was falling, she returned with her husband, Aristide Québriac, the Administrator-in-Chief of the Port Authority, who was also one of the leaders of the Resistance in the area. They took Cheney into the back yard and showed him a hiding place they had made behind a ceiling-high stack of fire-wood in a shed at the far side of the yard, accessed by a concealed door – a loose plank, which could be locked in place once he was inside. It would be his hiding place during any German searches.

He was desperate to discover the fate of his crewmates, but when the Resistance brought the first word of one of them, two days later, the news was bleak. A member of the Resistance had discovered a drowned body, dragged it from the water and hidden it from the Germans. Later that night one of the Resistance leaders woke Cheney and took him to identify the body. It was his wireless operator, Reg Pool. 'He'd been in the water two days,' Cheney says, 'and it was a really upsetting sight to see my friend this way. The Germans had refused permission for any burial but later the Resistance held a short service – a secret ceremony – for Reg which I attended. It was very sad but I was so pleased to be able to say goodbye properly and see him off with dignity.' The Resistance then buried the body in the local churchyard.

The body of the navigator, Roy Welch, came ashore some time afterwards, but both he and Pool had been badly wounded, and it was a third death that Cheney found hardest to bear. Rear gunner Noel Wait had been uninjured when he baled out of the doomed aircraft, but he was a non-swimmer and known to be terrified about baling out over the sea. His drowned body was found, still entangled in the shrouds of his parachute, and it is probable that his panic as he entered the water had cost him his life.

★   ★   ★

Back in England, 617 Squadron had lost another two members on 7 August, but this time they were killed in a training accident, which made the loss seem even more pointless and tragic. A Canadian pilot, Flying Officer Bill Duffy, and Phil Ingelby, Nick Knilans' best friend on the squadron, were practising target marking by dive-bombing targets on the Wainfleet bombing range in the Wash, and when Duffy pulled up from low level, one wing folded back and the Mosquito crashed into the water, killing both men instantly. Don Bell normally flew with Duffy, but the repeated dive-bombing practice made his stomach churn so much that he opted to stay behind on this occasion, saying that he intended to eat lunch for once instead, and Ingelby had taken his place, with fatal results.[3]

The raid on Brest on 5 August 1944 was also John Bell's last op with the squadron. He had completed fifty ops before standing down and was 'aware that the odds were getting a bit shorter of making it through,' he says. 'The flak over the targets was heavy and we'd lost a few crews, which concentrated the mind. Florence knew the risks too, and she would remind me from time to time that we needed to think about our future together and not to push my luck, to think of my responsibilities, of life after ops.'

★   ★   ★

As Don Cheney remained in hiding in northern France, 617 Squadron carried out another series of raids against the U-boat pens at Brest and the derelict French cruiser *Gueydon*, which the Germans were trying to move to act as a blockship to prevent Allied shipping from gaining access to the harbour. There had been little flak on the first raid, but as they approached the target area on 14 August, now looking more than a little scarred by the previous bombing, 'all hell let loose,' John Pryor recalls. 'They were ready for us this time. Heavy flak everywhere.'[4] As one of Willie Tait's crewmen also observed, Brest was definitely 'not a pleasant target' on this occasion. 'The daylight sky would be full of those large fourteen- and sixteen-inch naval shells that gave out a horrible orange cloud and a devil of a lot of shrapnel too. Also, Willie had an annoying habit of having a last look round that was not all that popular with the crew. However, we came through okay, although every machine in the squadron was hit in some way or another.'[5]

As John Pryor began his bombing run, his aircraft was hit by a heavy barrage of flak, knocking out the compass. The bomb-aimer, Pes Pesme, had identified the target, but there was then no further word from him. After trying and failing to raise him on the intercom, Pryor asked one of the other crewmen to go down and check on him. Pesme was still at his position but was stone dead, hit in the throat by a piece of shrapnel which had exited through the top of his scalp, killing him instantly. The bombsight had been running from the moment he was hit and, even though he was already dead, the bomb had still released automatically. There was nothing his crewmates could now do but turn for home. The homeward journey was, Pryor says, 'the longest, saddest and quietest (as far as talking was concerned) we ever had to make. I could not believe Pes was killed. It just would not register.'

All of them were still 'a bit shocked and stunned' when they landed back at base at 12.40. They were directed away from the normal dispersal to an open area where an ambulance could collect the body, while the ground staff began the gruesome task of cleaning out the bomb-aimer's compartment. Pryor and one of his crewmates wanted to go to Pesme's funeral to say a proper farewell to their comrade, but to their distress, they were told that it was against Group policy for crewmates of a dead airman to attend his funeral, and other members of the squadron had to stand in for them.[6]

★   ★   ★

Nick Knilans and his crew had been granted another week's leave in mid-August, and while his men went back to their homes or to stay with their English girlfriends, he 'went off to the fun and games of old London town for a farewell visit. The bartenders, owners and bar girls of the Chez Cup and the Chez Moi seemed to be glad that I was still alive.'

While he was on leave, Operation Dragoon, beginning on 15 August 1944, saw the first Allied landings in southern France. German counter-attacks were repulsed and by the night of 17 August, hotly pursued by the Allies and hindered at every step by the Maquis, the Germans were in full-scale retreat. On 19 August, heartened by the rout of the German forces, the French Resistance rose up in rebellion in Paris, and within a week the city had been liberated, with its German military rulers refusing to carry out Hitler's order to burn Paris in reprisal.

Nick Knilans returned from his leave to discover that, in his absence, 617 Squadron had been repeatedly bombing the U-boat pens at Brest and La Pallice. They had bombed every morning, and each time approached from the same direction off the sea. The flak gunners were therefore ready and waiting for the raiders when they appeared, and there was also a widespread and possibly well-founded belief among aircrew that German flak gunners gave the lead aircraft in a formation particular attention, believing that it would contain the highest-ranking officer. One pilot of relatively low rank but with a high number of combat ops had complained to Willie Tait about having to lead the formation every day, and Nick Knilans had volunteered to lead the squadron into combat in his place. However, 'higher authorities could not tolerate a Yank leading a Royal Air Force squadron into action', and his offer was refused.

After his return, Knilans, in company with his twenty-year-old flight engineer, Ken Ryall, who'd been made the squadron's Flight Engineer Leader, began a series of precautionary landings, heavy-weight take-offs and fuel-consumption tests, assessing what combination of pitch and throttle settings would consume the least fuel. They were accompanied by a civilian engineer from the Lancaster's manufacturers who gave them 'helpful hints on engine care and fuel consumption'. The reason for these tests was not revealed at the time, but it was clear to everyone on the squadron that they were preparing for an attack on an as yet unknown but far-distant target.

As Knilans' Lancaster 'droned at one thousand feet up and down the nearby coast', carrying out the repetitive tests, to his alarm he discovered that he was having trouble flying the aircraft, though not because of any mechanical problems. 'The low, broken clouds were casting fast-moving shadows on the ground. As I flew in and out of the clouds, I was becoming disoriented. I began to fly erratically.' He had already completed two full tours of combat operations and was well into his third, and he guessed that his subconscious was beginning to rebel against the continued strain on his nerves. As a result, fearing that he was endangering not only himself but his crew members as well, he had made the decision that the op for which they were training would be his last bombing trip for the Royal Air Force.

The next day he was given cause to regret not having stopped there and then, when he found himself tasked with doing an 'all-up weight test' which involved trying to take off from Woodhall Spa's relatively short runways under a total load of 70,000 pounds in fuel, bombs and the aircraft's own weight – 5,000 pounds more than the Lancaster's designed maximum limit. Accelerating even as he swung onto the runway, he roared along it, but the aircraft remained stubbornly earthbound, even as the end of the runway approached. He hauled it off the ground, but it sank back down, and he had to 'bounce' it off the ground again, by which time he was already 100 yards beyond the end of the runway. The control column felt 'flabby' in his hands and, only just above stalling speed and on a collision course with a line of telegraph poles and wires, he made the mistake of asking his flight engineer if he should try to fly over or under the wires.

'Oh bloody hell, do something quick!' was the only reply.

Knilans put down more flap, pushed the stick forward and they scraped over the top of the wires. He had to put the nose down again to pick up speed and stop the aircraft from stalling, then slowly crept higher, 'milking off' the flaps in stages. When he landed again, one of the civilian engineers who'd been watching from the control tower told Knilans he could not believe that a Lancaster could have taken off carrying that amount of weight from such a short runway. 'I told Wingco Willie Tait I would not recommend doing it again either,' Knilans said.

★   ★   ★

August 1944 had seen so many of Knilans' friends killed, wounded or posted missing that 'I was glad to see it end,' he remembers. One of the missing, Don Cheney, shot down over Brest on 5 August, was still in hiding, shielded by Aristide Québriac, his brave wife and the Resistance fighters and their supporters. 'They were all risking their lives to help me,' Cheney says. 'If we were ever caught, it was a PoW camp for me. They and their families would have been shot. They took huge risks to help me and without them I would have been a goner.' The local priest visited him every two or three days, bringing fresh fruit, books in English and a paper bag full of loose French cigarettes that the priest had cadged from his parishioners.

As American troops advanced close to the town, the German garrison withdrew and set up positions in the surrounding countryside. Although there was always the threat that they might return, Cheney was able to move around with a little more freedom. After seventeen days, on 22 August 1944, Cheney, in company with an American airman the Resistance had also been sheltering, was taken to Quimper.

German troops still occupied the city, though, says Cheney: 'they were hustling out of Quimper city hall, desperate to get out of the area as they must have realised that their time was up. I looked at their jackboots and their uniforms – here was the enemy on the run. But the dangers were still very real – who knows what they would do if they found a downed Allied flier in their midst? There was still a long way to go to safety.'

The two airmen were taken to a farmhouse outside the town for the night and then, still escorted by Resistance fighters, they went in search of the American lines. The battlefront was still very fluid and they never knew whether they were going to encounter German or American troops until they finally made contact with General Patton's Fifth Armored Division, who were bivouacked in a pear orchard. Any hopes that they had found safety were rudely dispelled by an American colonel who told them, 'Jesus, I don't want anything to do with you fellows.' He called in his Military Police and told them, 'Get these guys out of here.'

They hurriedly got back in the vehicle and drove off, arriving in St-Brieuc later that afternoon. There they were introduced to a Canadian intelligence officer who had been working undercover with the French Resistance, running an escape line for Allied fliers. He put them on the only freight train heading north that day, making for the city of Rennes, which was now firmly in American hands. That proved to be good news for Cheney's American companion, who was flown out the following afternoon, but the Americans wanted no part of Cheney, who was

dumped at the roadside outside the USAAF base and told to try
to flag down a passing RAF truck. No trucks came and on a
burning-hot day Cheney was 'eating dust all day as I walked back
into Rennes'.

As he passed through the city, he saw a line of women collab-
orators having their heads shaved by the Resistance and the
watching crowds scattering as German snipers still holed up in
the steeple of the church loosed off a few shots. A Frenchman
pointed Cheney to a road where he said there would be more
trucks, and after walking for about a mile he flagged down an
RAF pick-up that took him to Bayeux, where Allied 'walkers-
back' – downed airmen who had evaded the Germans with the
help of the Resistance – were being processed and repatriated.

When Cheney arrived, thinking his ordeal was over, he was
promptly placed under arrest, locked in a trailer for several hours
and then interrogated by a 'ramrod-stiff Guards colonel from the
old school, with a waxed moustache'. The colonel bombarded
Cheney with questions about his unit, his aircraft, the names of his
crew and a wealth of other details. He was then locked in the
trailer again for the night, while his story was checked with Military Intelligence in London. The next morning the colonel freed
him, telling Cheney his identity had been confirmed and he was
being sent back to Northolt.

'Thank you, sir, for your hospitality,' Cheney said. 'And thank
you for not shooting me as a spy.' The colonel never even cracked
a smile.

When Cheney flew in to Northolt, it was, he remembers,
'a wonderful feeling, almost too much to cope with, too much
to believe. We were given new uniforms, a good bath, and
deloused.' His first action was to send a telegram to his parents,
telling them he was safe – the only contact possible because they
didn't have a telephone – 'but in the midst of my joy and relief,'
he says, 'I was desperate to find out about my crew and what
happened to them.'

He had been taken to the old Madame Tussaud's building,
which was being used as a debriefing centre, and was coming

down the stairs when he bumped into his mid-upper gunner, Ken Porter, and his flight engineer, Jim Rosher, who had also been rescued and sheltered by the Resistance and had themselves arrived back in England only twenty-four hours earlier. Bearing in mind their near-death experience and astonishing escape from the enemy, their reunion seemed muted: 'We just hugged, chatted and tried to catch up with everything that had happened,' Cheney says. The other survivor, Len Curtis, had been captured by the Germans, but was freed when American troops overran the area and also repatriated.

Don Cheney reunited with his family

Although the fighting in Europe continued, Don Cheney's personal war was over, and he returned to Canada on Thanksgiving Day, 9 October 1944, two years to the day since he had left. His family and his girlfriend Gladys met him at the Ottawa Union railroad station. When he'd been shot down, his parents had received the news he was missing and had told Gladys at once. 'It was such a great shock,' she says, 'and then there was no more news, nothing at all. All I could do was sit and wait and hope. It was all about waiting: waiting for letters, waiting for telegrams, hoping for good news. Just waiting. Suddenly, there he was, running past the ticket collector.'

Cheney recalls:

I came through the barrier, and there were my mother, father and Gladys all lined up waiting. It was overwhelming for us all. They'd been there when I'd left two years before, now they were standing in the same place to welcome me home. I hadn't even spoken to any of them for two years, but here was the image of home I had held on to and longed for during the darkest times. I realised how lucky I'd been. It was difficult to believe everything I'd gone through and yet I was now home. It was almost unreal. The first thing I asked for was a banana; I hadn't seen one for years and really wanted one now! We piled into the car, laughing and crying, it was all a blur and such a happy time, but tinged with real sadness for my dead crew. Inside, I was still crying because of them and I still think of them every day.

He and Gladys were married six months later, on 21 April 1945.

★   ★   ★

On 1 September 1944, Nick Knilans dropped his 130th practice bomb. 'I tried to add up the weight of all the bombs I had dropped on enemy cities and targets,' he said. 'It came to seven hundred and fifty thousand pounds. The number of human beings I had killed or wounded, I did not want to think about.'

He continued to fly more fuel-consumption tests and to practise 'short-landings', when, by 'dragging' the Lancaster in at low speed and low level, he was able to land and bring it to a halt in less than 800 yards instead of the normal 1,200–1,600. 'Unbeknownst to me, this practice was soon to stand me in good stead.'

The true purpose of all this practice was soon to be revealed at a briefing for an operation that, if successful, was guaranteed to generate almost as much publicity and propaganda as the Dams raid that had made the squadron world-famous.

CHAPTER 11

The Beast

Tirpitz *firing its guns*

On 28 August 1944, some of the 617 Squadron aircrew were playing football on the airfield at Woodhall Spa when a stranger appeared on the touchline and asked for their CO, Wing Commander Willie Tait. Some time later, Tait returned to tell the footballers they had 'a special job' but would offer no other details until the day of the op. So it was not until 11 September 1943 that the crews made their way to the briefing room at Woodhall Spa, to be greeted by Tait, who was to brief them on Operation Paravane.

By now German forces were in full retreat across Europe and the tally of Allied victories and cities liberated was mounting by the day. German forces had surrendered at Toulon and Marseille on 28 August, Dieppe fell to the Allies on 1 September, Brussels was liberated two days later, Antwerp two days after that and Ostend on 8 August. On 10 August, Allied forces advancing from Normandy and from southern France linked up at Dijon, cutting

France in half. That same day, Allied troops set foot on German territory for the first time as they advanced into Aachen.

The atmosphere as 617 Squadron assembled for their briefing the next day was a world away from the dark days when the squadron had first been formed, but they knew that there was still much to do both to guarantee the ultimate victory, and to minimise the Allied casualties it would entail. As they entered the foyer, the first thing they saw was Barnes Wallis standing proudly alongside one of his Tallboys, resting on its vanes, upside down.[1] It was, said Bruce Buckham, 'a beautiful big bomb [and so] highly machined, you could have shaved in it'.[2]

Until then, the crews had been given no clue about their target, though 'we knew it was a long way away,' Iveson said, 'because we were practising cross-country flying with different path settings and different altitudes to work out fuel consumption. We knew it was going to be over a lot of water because we did dinghy drill. We knew it was a high-level attack because we did a lot of high-level bombing practice.' They had been told only to bring 'small kit' – toothbrush and shaving tackle – with them, leading some to assume they were being sent to attack targets in Italy and then flying on to North Africa to refuel, as they had done before on two or three occasions.[3] When the aircrews entered the briefing room, they saw that Tait had something on the table in front of him, shrouded with a black cloth. Like a conjuror producing a rabbit from a hat, he allowed the tension to build for a few moments and then whisked the cloth away with a flourish, revealing a scale model of the German battleship *Tirpitz*.

Launched at Wilhelmshaven in 1939, *Tirpitz* was even bigger than its sister ship the *Bismarck*, and was the largest and most modern German warship ever built. Two hundred and fifty metres long and displacing over 50,000 metric tons when fully loaded, it was so big that novice seamen often got lost as they tried to find their way around. It had its own library, post office, cinema, bakery, and even published a twice-weekly newspaper for its 1,700-man crew. It could cruise at 35 miles per hour, faster than any British warship, its battery of 120 guns included eight 15-inch guns with a range of 17 miles that could menace targets invisible beyond the horizon, and, like all the *Tirpitz*'s guns, could be elevated enough to target incoming aircraft. The ship was so heavily armoured that conventional bombs simply bounced off a toughened steel carapace that was over a foot thick.

Hitler called it 'the pride of the German Navy'. Churchill growled that it was 'The Beast' and, based in occupied Norway, it was a constant potential threat to the Arctic and Atlantic convoys. 'I crossed the Atlantic on the *Queen Elizabeth* with six-teen thousand American troops,' Tony Iveson said, 'and there was still an awful lot of transatlantic traffic: personnel, weapons, food and fuel. If the *Tirpitz* had gone into the Atlantic, sunk one or two of those large ships, drowning a lot of American troops, it might have affected America's attitude towards Europe and the war.' The mere existence of the *Tirpitz* was sometimes enough to cause panic. In July 1942, false rumours that the battleship was at sea had persuaded the Admiralty to order the convoy PQ17 to scatter, and without their escorts to protect them, twenty-four of the convoy's thirty-five ships were lost.

There had already been thirty-three previous attacks on the *Tirpitz*. The Royal Navy had sent five aircraft carriers to destroy it, but despite attacks by hundreds of Fleet Air Arm aircraft, midget submarines – all of which were destroyed in their attack – torpedoes and dive bombers, the *Tirpitz* remained afloat. Now the Royal Air Force's 617 Squadron had been given the task of finishing the job.

Although British commanders clearly regarded the *Tirpitz* as a vital target – and three British capital ships were kept tied up in Scapa Flow just in case the *Tirpitz* emerged onto the open seas – to at least one crew member of 617 Squadron, the battleship was 'just another target. Nothing special.'[4] Thirty-six Lancasters were to take part, eighteen from 617 and another eighteen from 9 Squadron at Bardney, plus a Lancaster photography plane, 'Whoa Bessie', from 463 Squadron – one of four all-Australian RAAF squadrons attached to Bomber Command. Captained by Flight Lieutenant Bruce 'Buck' Buckham, his aircraft would be carrying four extra men in addition to his crew: two cine-cameramen, BBC correspondent Guy Byam and W. E. West of the Associated Press.

Buckham was a twenty-six-year-old who had completed his first thirty-op tour of duty on bombers in early 1944. It had included a solo mission to Berlin and a raid on the Krupp steel-works in Essen, during which he lost two engines and, hit by flak, had to extinguish a ferocious on-board fire by throwing his bomber into a perilous dive, earning the Distinguished Flying Cross for his valour.[5] Recalled for a second tour, Buckham and his crew were then selected to fly a special Lancaster fitted with camera equipment to record high-profile bombing operations at close range.

The Lancaster bombers were perfect for the task: 'The Lancaster was noisy and could be uncomfortable, but it was totally and utterly reliable'[6] and 'a damn good bombing platform. It was very, very steady and there was no better launching pad ... particularly on a precision bombing such as the *Tirpitz*.'[7] The Tallboy bombs they were carrying were 'an incredible sight', one crewman said, 'something to be reckoned with, an amazing piece of engineering which almost filled the bombbay,'[8] and they were the only RAF weapons capable of penetrating the battleship's thick steel armour plating.

Intelligence reports indicated that, protected by submarine and torpedo nets, the *Tirpitz* was anchored in Kaa Fjord, an arm of Altenfjord (now Alta Fjord), protected by a wall of mountains, inside the Arctic Circle, right at the northernmost tip of Norway. With a Tallboy in the bomb bay and full fuel tanks, the Lancasters weighed over 67,000 pounds at take-off, and the *Tirpitz's* location put it beyond their range for a return flight, even from the closest UK base in the north of Scotland.

The only option was to break the trip into stages, refuelling in northern Scotland and then flying on to a base near Archangel – a 'city', actually little more than a huddle of wooden buildings, in Soviet Russia, several hundred miles south-east of the target. That would serve as the base from which to attack the *Tirpitz*, returning to Russia to refuel again before flying back to Britain. Even using Archangel as a staging post would necessitate a twelve-hour flight from Britain. 'It would have been nice if they had added a co-pilot,' Nick Knilans said, with a smile. 'I did not look forward to a twelve-hour flight without any help. I did not expect to leave my seat in the cockpit either. Nor did I.'

In order to have the range to reach Archangel, the Lancasters also had to be stripped of their mid-upper turrets and fitted with overload tanks in the fuselage above the bomb-bay. Knilans and his flight engineer, Ken Ryall, briefed the others on the optimal pitch and throttle settings to make the most of their available fuel. Two Liberator aircraft were also being used to ferry ground crew, spares and all the other equipment needed to keep thirty-seven Lancasters combat-ready, and the station's medical officer went with them to tend to any illness victims or casualties.

The briefing on 11 September 1944 made use of a large-scale model of the fjord, including the *Tirpitz*, anchored close to the wall of mountains that made any airborne attack incredibly difficult.[9] In addition to the ship's own guns, there were flak batteries on the shores, and a pipeline with a series of smoke canisters encircling the fjord, able to create an almost impenetrable smoke-screen within ten minutes of a warning being given. Tait hoped that by approaching at low level and from an unexpected direction – the south or south-east – he and his men would give *Tirpitz* no more than eight minutes' warning of their approach, leaving a two-minute window in which to drop their bombs before the smokescreen made accurate bombing almost impossible. But to take advantage of that two-minute window of opportunity would require almost twenty-seven and a half hours of total flying time.

Later on 11 September – Tony Iveson's twenty-fifth birthday – the Lancasters thundered into the air from Woodhall Spa and, having refuelled at Lossiemouth, set course for Russia. One aircraft from 9 Squadron had to be left behind, after its Tallboy slipped off its mountings in mid-air and had to be jettisoned. The rest, carrying a ton more weight in fuel and bomb-load than the

Lancaster's design limit, flew past the Orkneys and the Shetlands and across the North Sea at low level – 400 feet – to avoid detection by German radar, skimming over a few isolated fishing boats. The navigators plotted the course by dead reckoning, and 'it required great concentration to keep track of where we were,' Basil Fish, one of the 617 Squadron's navigators, says, 'and where we were supposed to end up. Quite a journey into the unknown.'[10]

As they reached the Norwegian coast, they could see the fjords and mountains by the light of the half-moon. Climbing to 14,000 feet, they cleared the peaks then crossed into Sweden and the Gulf of Bothnia and turned north. Malcolm 'Mac' Hamilton's crew 'got all excited' looking down from the windows at a brightly lit town below them – they'd never seen lights from the air before. 'We'd been under the blackout in Britain and across the whole of France and Germany. Someone said, "It's like fairy-land, isn't it!" Then we were back into the dark as we flew on into Finland.'[11]

As they flew on, they could see vast swathes of blackened forest that had been destroyed in the Russo-Finnish War of 1940. Even though the Russians had given the squadron a call sign, to identify themselves as friendly forces, Soviet guns fired at them as they crossed the border into Russia. Hamilton's gunner, Gerry Wither-ick, 'a Cockney through and through, and a great humourist',[12] was so furious at being fired on by supposedly friendly forces that he claimed to have written 'G A Witherick' right across one of their gun emplacements with his own guns.[13]

Witherick did 100 ops as an air gunner and always claimed, with a wink, that it was the easiest way he had ever found to earn a living.[14] He had also flown numerous anti-submarine patrols with 405 Squadron, but claimed that he never counted them in his total of ops as 'all we did was to fly over water!' He added that while he never saw a single enemy submarine, he was shot at by the Royal Navy every time he went out to help them.[15]

During the marathon flight, several Lancasters had also come under anti-aircraft fire from Swedish and Finnish guns, and Flying Officer Bill Carey's Lancaster caught so much flak that, after he landed, he took no further part in the op. It was not the last flak attack Carey would endure.

As dawn broke, the aircrews found themselves over the 'vast forests and swamps which characterise that part of Russia'.[16] The weather had been poor all the way and there was no let-up as they approached Archangel. The Russians had assured them that there was perfect visibility, but in fact, Bruce Buckham said, 'it was ten/tenths the other way: low cloud, raining, pretty well down to the deck'. They were flying in low cloud just over the tops of trees with mist laced between them, and even at that very low level the aircrew had only occasional glimpses of the ground below. Larry Curtis described the maps that the navigators were having to work with as 'pretty poor. The signals information was pitiful.'

The navigators were plotting the course by dead reckoning. 'We had no radio beams or beacons to guide us,' Nick Knilans said, and their only maps were outdated relics from the British North Russia Expeditionary Force's occupation of Murmansk and Archangel in 1918–1919, in a futile attempt, in its chief architect Winston Churchill's words, 'to strangle at birth the Bolshevik State'. 'Now you have got radar, GPS, and all these things, you could go and find Archangel,' Tony Iveson said. 'Frankly we had Sweet Fanny Adams, beyond the co-ordinates of the airfield we were going to.' The Russians were supposed to be sending out a radio signal for the Lancaster crews to home on, but their radios remained silent. 'I think someone forgot the difference between the Russian and English alphabet,' Tony Iveson said, 'so when the wireless operator tried to make contact, he got nothing.'

As a result, as Bob Knights wryly commented, navigation 'was all really by guess and by God', and given the lack of maps and radio communications, and the appalling weather conditions, it was thanks to a truly astonishing feat of flying and navigation that they found Archangel at all. Bruce Buckham descended through the dense cloud to Onega Bay on the White Sea about 125 miles west of Archangel, where he found eight or ten other aircraft 'milling around'. They then flew 'like a string of ducks' about 50 feet above the beach right around the foreshore to

Archangel and then down the river Dvina looking for Yagodnik aerodrome, where they were supposed to land. It was a grass airfield on an island in the middle of the river, but the weather was so bad and visibility so poor that many of them couldn't find it. Almost all the pilots had similar difficulties, even when flying as low as 100 feet, and some of those who did land had branches and foliage from the treetops they had clipped jammed in tail wheels and elevator controls.

The flight was 'at the limits of endurance', Tony Iveson said. 'I was in the air for twelve hours and twenty minutes; I had never done anything like that before. To some degree we were all exhausted, but when you're not quite sure what's going to happen, you maintain a certain awareness!'

Everyone was running short of fuel, and in desperation Bruce Buckham led some of his 'flight of ducks' back to the river mouth, where they landed at a tiny airfield on Kergostrov Island. It was normally used by aircraft no larger than two-seater Tiger Moth trainers, but somehow the monster Lancasters with their 12,000-pound bomb-loads managed to land there, though when Buckham did so, 'some incongruous sights met our eyes. There were Lancasters with their noses pushed into buildings, some with only their tails showing through walls or roofs, and others on their bellies.' A few of those had to be declared 'u/s' – unserviceable – and abandoned there.

Mac Hamilton eventually found Yagodnik and did a run across the airfield, but couldn't see any markers, though he saw a wind-sock at one end, confirming it was an airfield of sorts, albeit a very rough one. 'I looked down at what I thought were posts with white tops on them.' When he got a bit lower he realised that they were Russian sailors wearing white hats – 'they'd marked it [the runway] with their own men!'[17]

Those who reached Yagodnik were greeted by a Russian band and a banner reading 'Welcome to the Glorious Fliers of the Royal Air Force'. However, only about half the Lancasters made it; the rest of the 'glorious fliers' were 'spread out over Russia' or had crash-landed – six pilots had to make emergency landings. Flying

Officer Ian Ross searched for Yagodnik for two hours and then, with only 30 gallons of fuel remaining, opted to crash-land, wheels-up, in a marshy field alongside the railway to Molotovsk. He put down safely without injury to the crew, but his Tallboy was torn loose and 'ploughed a furrow about a hundred yards long'. By the time the crew clambered out they were confronted by 'six Russian soldiers with machine-guns, one of whom was kicking the nose of our bomb and another trying to steal the bombsight'.[18]

A packet of English cigarettes was enough to restore international relations, but the aircraft was a write-off, as was Squadron Leader Drew 'Duke' Wyness's Lancaster. He landed in a grassy field apparently used by Soviet fighters, but the ground was so wet that when he applied his brakes the plane simply skidded across the field and slid into a fence, wrecking the port undercarriage. Four 9 Squadron aircraft were also damaged beyond repair.

Nick Knilans had kept descending as the cloud base lowered. Finally he was 'at treetop level and the clouds were there too'. Larry Curtis, Knilans' wireless operator, knew that they were within 60 miles of Archangel but 'frankly got lost ... We had a twelve-thousand-pound Tallboy bomb underneath, we were pretty lost, the weather was awful.' They descended as low as Knilans dared, trying to pick up a landmark, but all they could see were rivers and pine forest. Curtis remembered:

> We had decided to ditch in one of the rivers, and had got rid of the escape hatches. All the target maps went out of the window, my brand-new hat which I'd bought to impress the Russians went out. We were just about to ditch when we saw another Lancaster. We presumed he knew where he was so we flew up to him. He probably thought the same about us, so we went in circles.[19]

Eventually, desperately short of fuel and still with no accurate idea of their location, Knilans turned on to a reciprocal course and flew back to the White Sea, where he glimpsed a hayfield on the edge of Onega Bay that looked as if it might once have been used as an airstrip. As he was circling over it, he saw 'five barefoot boys about ten years old. They made vigorous motions with their heads and hands. I gathered that they were signalling that it would be OK to land there.' He flew up a gully below the level of the field, with the smoke from a house chimney showing that he was approaching downwind, but he 'did not expect to float far with a six-ton bomb aboard'.

He pulled up as he reached the barbed-wire fence surrounding the field, then cut the throttles the instant he cleared it, landing so close to it that the rear gunner later told him that his gun barrels had clipped the barbed wire. He couldn't keep the brakes on because mud and turf were piling up in front of the wheels, threatening to break the undercarriage, but the far fence was rapidly approaching. He managed to slew around a haystack and came to a halt next to the boys without any major damage to his aircraft. He gave them a smile and thumbs-up sign by way of thanks. He had been flying for twelve and a half hours, and was so short of fuel that two of his engines had cut out on landing.

Tony Iveson had seen him land and now came in himself, 'empty and upwind', but landed safely. 'I felt to press on blindly into the unknown with fuel draining away was mad,' Iveson said. Soviet troops eventually appeared. 'They were wearing grey-coats, hadn't shaved for a week and looked as if they would cut our throats for a rouble or two.' They put the aircrew in the back of an open truck and jolted over a rough track into the town, where they met the Town Major. With neither side able to communicate with the other in English, Russian or the air-crews' smattering of French or German, they were taken to a loghouse and locked up while their hosts radioed Archangel or Moscow for instructions.

An hour later, having established that the aircrew were friends and allies, they released them and fed them salmon, potatoes and black bread and what appeared to be glasses of water, though as Iveson discovered when he downed his in one, it was actually vodka. 'The Russian Major stood up and toasted the Royal Air Force,' he said, 'so I stood up and toasted the Red Army.'

Although East–West relations had now thawed a little, a soldier armed with a rifle with fixed bayonet still stood guard over them and accompanied them whenever anyone wanted the latrine. There was no basin, nor even any water, only a log raised knee-high above an open ditch. 'I thought that they could have peeled the bark off the log at least,' Knilans later joked. They discovered they were being housed in a rest camp for survivors of the Battle of Stalingrad, but 'it was pretty grim,' Larry Curtis said, 'and seemed to have been built in a swamp; most of the paths were trees that had been cut down. It was the first time I saw Red Army women. They did the same jobs and looked more like men than the men, but they were kind and helpful.'[20]

The Russians succeeded in manhandling the two aircraft around – a task that would have been easier had they remembered to take the brakes off first – and lined them up, using the main street – which was surfaced with logs laid side to side – to serve as a rough and very uneven runway. They even extended it by laying more logs while the aircrews were sleeping.[21]

A Lancaster forced to crash-land searching for Archangel airfield

The next morning Willie Tait arrived from Yagodnik in an antique Russian biplane and summoned a transport aircraft, bringing just enough fuel in 50-gallon drums to get the Lancasters to Yagodnik. The fuel all had to be laboriously hand-pumped into the Lancasters' tanks and, said Knilans, 'the Russians let us do all the pumping'.

Iveson and Knilans then took off again. Iveson had a relatively trouble-free flight but Knilans and his crew were lucky to survive. With the Merlin engines on full emergency power, he managed to get airborne right at the water's edge, and climbed to 1,000 feet. However, the long flight to Russia on a lean fuel mixture had caused the spark plugs to foul up, and when he throttled back to keep behind Tait's slower aircraft which was leading them to Yagodnik, his engines coughed, spluttered and then stopped. He shoved the control column fully forward with his left hand and simultaneously rammed all four throttle levers to the stops. The increased air pressure from the dive turned the propellers over enough to restart the engines, and he began to pull up from the dive just above the forest canopy.

Wanting to help, the flight engineer raised the flaps but, never having touched them before, he did not realise that this would cause the aircraft to sink or 'mush' several feet. The aircraft lost enough altitude to drop into the tops of the trees, cutting a swathe through them like a sickle through grass. Knilans could not pull up more sharply, since that would have dropped the tail even deeper into the trees, with fatal consequences. However, the thicker trunk of a lone pine tree towering above the others was now looming ahead of them. Knowing that it would wreck the propellers if it hit them, but unable to avoid it completely, Knilans aimed straight for it, taking the force of the impact on the Perspex nose of the aircraft.

A three-foot section of the tree was driven through the bomb-aimer's Perspex 'goldfish bowl', narrowly missing Knilans and his bomb-aimer and flight engineer. The impact also smashed the Perspex cockpit, destroyed the bombsight and tore off the doors of the bomb-bay. The slipstream roared through the broken cockpit and out through the dinghy escape hatch over the pilot's head, and everything movable that had not been lost or jettisoned in the landing the previous day, including navigator Don Bell's maps, was now sucked out of the aircraft by the ferocious wind.

'They all shot up through the escape hatch to flutter about over the Russian landscape.' The starboard engine's radiator had also become so clogged by pine cones and needles that it overheated and had to be shut down.

Knilans had to fly on one-handed, covering his eyes with his other hand to protect them from the ferocious wind and squinting out through slightly separated fingers. Even though some of his controls were jammed, he managed to land at Yagodnik on three engines. As they clambered out of the aircraft and rushed across the airfield to join the briefing for the attack on the *Tirpitz*, Don Bell still had pine needles and bits of twig in his hair.[22] The piece of the tree was retrieved from the aircraft and later became a treasured souvenir in the Mess at Woodhall, captioned 'Believe it or not!'[23]

Collisions with the buildings surrounding the Kergostrov airfield had already damaged a few Lancasters when they landed there, and they proved equally dangerous when they took off again after the weather cleared a little. Bruce Buckham was the first to take off towards the buildings at the end of the runway. He 'literally bounced the aircraft into the air by moving the control column backwards and forwards and getting a big enough bounce, and at the appropriate time, turning it on its side and flying sideways out through [a gap between] these buildings.'

All he could hear over the RT was the other pilots saying, 'Oh Christ! Oh shit! Have we got to do that?'

'Yeah,' Buckham said. 'That's the only way you're going to get off.'[24]

Once airborne – by a miracle without further aircraft losses – Buckham's 'flight of ducks' made the short flight down to Yagodnik, where the surviving aircraft were being readied for the attack. During the day most of the other missing Lancasters arrived, but only twenty-seven of the original force of thirty-seven aircraft were still serviceable as, in addition to the write-offs, the ground crew had to cannibalise some damaged aircraft to repair the others. Refuelling was also a painstaking business, again using hand-pumps to transfer the fuel, this time from 44-gallon drums. It took over eighteen hours to refuel the entire force.

With thick cloud cover over the target, the crews were stood down until the skies cleared, and they remained at Yagodnik for three days. Each morning the PRU (Photographic Reconnaissance Unit) Mosquito took off and overflew the target area, to check that the *Tirpitz* was still anchored there and to look for a break in the weather that would allow the attack to be launched. The Mosquito, painted pale blue, was much admired by the Russians, who had never seen such a beautiful aircraft before.

While they waited for the skies over the target to clear, Knilans and some of the other 617 Squadron aircrew were billeted in an underground barracks with a wood stove at one end and a wash-room and bucket toilets at the other. The barracks was presided over at night by a 'large, pleasant-mannered female soldier'. Although they could not speak each other's language, Knilans 'gave her some body powder and cigarettes' and they bonded enough to take a walk in the starlight. 'There was a bit of Allied nuzzling and a mutually satisfying frolic on the tundra.'

The remaining aircrew had been billeted on the *Ivan Kalleyev*, a bug-infested river steamer anchored in the river Dvina. The Russians said it was the best accommodation there, though Bruce Buckham thought 'it made the *Altmark* [a German oil tanker and supply vessel] look like the *Queen Mary*'. 'The bugs bit us to bits,' Iveson said, smiling at the memory. 'The only man they didn't bite was the CO – even Communist bugs had respect for rank!' Bad as it was, the ferry boat was more salubrious than the accommodation provided for the ground crew who, like Knilans' crew, were given dank, stinking underground quarters that housed 'breeding colonies' of lice and bedbugs.

Despite the medical officer spraying their cabins and dusting their sheets with insecticide powder, by the next morning several of them were 'covered in enormous red blotches, one chap's eyes were completely closed up and another looked as if he'd developed mumps in the night.'[25] As at Onega Bay, 'women seemed to do all the work and were loading logs on the ship and ... women cleaners came in every morning to wash the floors and paid no attention to us getting up half-dressed. The Russian men stood on the deck smoking while the women worked.'[26]

The Russians entertained their visitors with vodka. It was not the familiar clear spirit, but 'a thick syrupy stuff like paraffin … We didn't perform too well and got quite sloshed,' Buckham recalled. One Canadian was sufficiently inebriated to decide that a swim in the freezing river was a good idea and had to be rescued by a Russian launch. The following night the visitors returned the hospitality. 'They were doing quite creditably,' Buckham said, 'until we got a bit naughty and started mixing whiskey with brandy and gin and everything – they were real Molotov cocktails!'

Even though the aircrew didn't have any kit with them, they also played a football match against the Russians. It was supposed to be a friendly, but some of the opposition were 'Mongols, about seven feet tall with stainless steel teeth – they bowled us over like ninepins!' The Russians won nine–nil, and when the 617 Squadron men weren't picking the ball out of their own net, they were standing to attention, because the game stopped after each goal while the Soviet national anthem was played.

That night, the Russians laid on a film show for their guests. The film was a record of the Nazi invasion of the Soviet Union and some of the atrocities they had committed, secretly filmed by concealed cameras. 'After witnessing that for half an hour, we'd had enough,' Buckham said, 'because it wasn't nice seeing people raped, others garrotted and strung up to lamp-posts, and so on. We got up to walk out, but we were very smartly jostled back with tommy guns pointing at our guts, so we had to sit down again and take the rest of it.'

They also had to endure the sight of other atrocities, this time perpetrated by their allies at much closer quarters. The steamer in which they were housed was normally used for transporting political prisoners up to the Lake Yagoda area, and the 617 Squadron aircrew saw many prisoners there, cutting up wood for the winter. 'As they ran out of steam, the NKVD man [guarding them] used to take a stock whip off his shoulder and flick them with the whip. They'd do that for a couple of days and when they didn't get up, they just pumped them full of bullets. This was right in front of our eyes and we just had to take it, we couldn't do anything about it.'[27]

Another film show was scheduled for the following night, but after protests to their interpreter, who had 'silver-filled teeth and smelled strongly of perfume', Popeye and Mickey Mouse cartoons were shown instead of more propaganda films. They also learned to dance the Russian way and taught the Russians to jitterbug in return.[28]

★   ★   ★

To the aircrews' relief, on 15 September 1944 the pilot of the PRU Mosquito at last flew low over the airfield and fired the green flare that showed the skies over Altenfjord were clear. The twenty-five remaining Lancasters, already fuelled and bombed up, took off at once, serenaded by a Russian band playing 'Hail the Conquering Hero Comes'. 'It seemed a bit premature in retrospect,' one aircrewman remarked.[29] They headed towards the target at low level, keeping below 1,000 feet to avoid detection by German radar. Among them, astonishingly, was Nick Knilans' Lancaster, which the ground crew had managed to restore to flyable condition.

The Russian fuel was not of the highest quality and purity, with the result that 'when you got the revs down and put a bit of pressure in the cylinders, the thing "pinked" like mad. So we had to keep the revs up, but we weren't too bothered about fuel as it was only a six-hour trip.'[30] The flight to the target crossed Russia, Finnish Lappland and Finnmark, but though they overflew a German airbase in Finnish Lappland, no fighters rose to intercept them. There was a fighter base at Bardufoss, south of Tromsø, but the Germans evidently considered the flak batteries and smoke-screen enough protection for the *Tirpitz*, and there was no fighter cover as the bombers approached. 'We didn't believe that they would reach us up there in the north,' one of the *Tirpitz*'s anti-aircraft gunners, Klaus Rohwedder, recalls, but such complacency was rudely shattered as the ship's alarms began to sound and the crew scrambled to battle stations.[31]

Although Bruce Buckham in the camera plane went in 'pretty low', the remaining aircraft made a steep climb to avoid light flak and reach their normal bombing height of 13,000 to 18,000 feet, to allow the Tallboy to reach its maximum terminal velocity. Tony Iveson had his first sight of the *Tirpitz* as he reached 16,000 feet: 'I just saw this black shape against the cliffs. From that height it didn't look like the awesome battleship that I had heard of, it looked like a toy.'

As they climbed, they could see from 15 miles out that the *Tirpitz* must have had an early warning, because the pipeline around the shore was already belching out thick black smoke. Had the crews flown straight in, they might have had a better sight of the target, but they kept to the pre-planned dog-leg approach, and by the time most were beginning their bomb-runs, 'literally hundreds of smoke canisters on the water, the ship, and the land' were doing their job so effectively that the fjord was practically full of smoke.

Knilans and a handful of others managed to drop their bombs before the ship was completely obscured, and his bomb-aimer 'Taffy' claimed a direct hit, though, as Knilans wrily acknowledged, Taffy 'never failed to claim a hit on target'. The anti-aircraft guns had opened up when the Lancasters were eight minutes from the target. 'There was a blue searchlight and then the gunfire,' one crewman recalled. 'We looked down and you could see *Tirpitz* under the cliffs ... but there was so much flak we didn't stand much chance of getting near it.'[32] Several bombers were hit, including that of the accident-prone Knilans, who lost an engine after dropping his bomb and had to make another three-engined landing, the seventh of his career, back at Yagodnik after the raid. The battleship's guns were still firing at the raiders, but that largely helped to improve the aiming point for the bomb-aimers. Despite the difficulties in clearly identifying the target through the smoke, seventeen Lancasters, led by Willie Tait, dropped their Tallboys.

A 9 Squadron pilot claimed the first hit at 10.55 that morning, 617's John Pryor claimed a second a minute later, and Squadron

Leader Fawke a third at nine minutes past eleven. As well as Knilans', there were three other probable near-misses by other 617 crews, and Wing Commander Tait's Tallboy was thought to have 'gone right through the bows of the vessel and exploded in deep water'. However, the aircrews' initial impression, apparently confirmed by subsequent surveillance flights, was that neither this nor any of the other bombs had caused fatal damage to the *Tirpitz*, and in any event the smoke and the flak bursts made the claims of hits or near-misses problematic. 'One of the lads I knew said he'd hit the *Tirpitz*,' Wilfred Bickley, a Devonian with a broad accent and an even broader smile, recalled with a laugh, 'but when they checked the picture, he'd only blown up the gas works about three miles away!'[33] Bruce Buckham dropped the eight 'Johnny Walker' magnetic mines that he was carrying inside the torpedo nets around the *Tirpitz*, hoping that their oscillating track when in water would draw them against the ship's steel hull, but once more there was no visible result.

Bob Knights' bomb-aimer couldn't see the target for the smoke and didn't bomb at all. Knights later recalled that 'Leonard [Cheshire] always said: "Barnes Wallis has produced these marvellous bombs, they're very expensive and I don't want them thrown all over Europe. If you can't put them on the target, bring 'em back!"' 'We were told that these new bombs cost £1,500 each,' Tony Iveson added, 'a huge amount of money then, so we didn't want to spray them around the countryside.'

Like Knilans, Knights lost an engine and landed at Archangel on the remaining three. Not only did they have to replace the faulty engine but the Russian ground crew dropped the Tallboy through the open bomb doors and onto the ground. Luckily for them and the watching British aircrew, 'it was all soggy grass' and the bomb did not detonate.

While the rest of the Lancasters returned to Yagodnik to refuel, Bruce Buckham's less fuel-hungry aircraft had begun the long flight direct to the UK. He first had to fly towards the North Pole, beyond the range of German fighters based in Norway, and then turned south, flying on over the North Sea 'through the most atrocious weather I'd ever seen ... There was nowhere to fly apart from the space between the tops of the angry sea and the bottom of the snow- and ice-laden cloud.' He battled through it for six hours, averaging only about 60 miles an hour instead of the Lancaster's normal 200 to 220 mph, and was desperately short of fuel when he at last landed back in Lincolnshire. He and his crew had been airborne for fifteen and a half hours, 'the longest bomber flight and certainly the longest by a Lancaster throughout the war'.[34]

On their return to Yagodnik, the remaining Lancaster crews had to wait for the reconnaissance Mosquito to come back to find out if the raid had been successful, though most already shared Tony Iveson's belief that 'We hadn't finished with *Tirpitz*. There was relief that we were off the target, but also the knowledge that we were going to have to go back and revisit her.' That was confirmed when the reconnaissance photographs showed that the *Tirpitz* was still afloat, 'which absolutely broke our hearts', Larry Curtis said.[35] Even worse, 'the Russians minced no words about our failure: They knew we wouldn't be able to sink the *Tirpitz*.'[36]

The Russians laid on some entertainment for 617 Squadron that night. They climbed aboard a boat, steamed downriver to Archangel and marched from the dock to an auditorium where a stage show had been put on. Mac Hamilton and Nick Knilans decided to make their own entertainment and slipped out of the darkened theatre. Having sold some cigarettes for roubles, they found a nearby hotel where they drank some straight shots of vodka in the bar. Before long their roubles were running out, but they then got chatting to two British sailors who took them to the locked building that held the ships' stores for Royal Navy vessels visiting Archangel. One of them had a key and five minutes later they were heading for an open peasant market, each with a gallon can of jam under his arm. They sold them to eager buyers, and, well equipped with a fresh supply of roubles, went back to the hotel bar, where some Russians joined them. Some time later, the two sailors were fast asleep with their heads on the table and two of the Russians had fallen off their chairs and were lying on the floor but, said Knilans, 'Mac and I were feeling no pain,' when a strange RAF Wing Commander appeared.

'Pardon me,' he said, 'but aren't you gentlemen part of the group from the auditorium? Well, the rest of your chaps are down on the dock. They've been there for about an hour by now.' He turned out to be the Air Attaché from the British Consulate in Archangel. He led them from the hotel to a waiting staff car and drove them to the dock, where 'Wingco Tait and the other air-crew did not seem too amused at our holding them up. I decided to stay out of wrath's way,' Knilans said, 'and went down below deck. The Russian captain gestured me into a small kitchen–dining room. The cook gave me a plate of fried fish and mashed potatoes. All in all, I was quite pleased with the day's events.'

On 17 September 1944, after their aircraft had been refuelled and made serviceable, Willie Tait led a formation of sixteen Lancasters out of Yagodnik for the long flight back to Woodhall Spa. Some aircraft were still being repaired and six others were write-offs after the crash-landings, so their crews distributed themselves among the other Lancasters, but one of the sixteen that flew out of Yagodnik that day never made it home. Flying Officer Frank Levy, a tall, dark and serious-minded Rhodesian pilot, inexplicably crashed near the summit of the Rukkedalen Mountains in Norway. He and his crew, together with two members of Squadron Leader Wyness's crew, who had been hitching a ride home with them, were all killed instantly.

'To this day I do not know how it happened,' Tony Iveson said many years later. 'Because the mountain was only three thousand five hundred feet high. I have never understood why they were in that area anyway – it was well off track – but what was wrong I have no idea.' When, sixty years after the war, Iveson paid his respects at the graveyard in Nesbyen where all of the men were buried, he found that their graves were still being tended by the villagers.

More Lancasters left Yagodnik on the two following days, the last of them flying out on 20 September. Nick Knilans had been among those who flew home on 17 September, and had he not already made the decision to stand down from ops, his serial mishaps during the *Tirpitz* raid would have served as a timely reminder of the need to do so. When he landed at Woodhall Spa after his last-ever flight with 617 Squadron, his battered Lancaster was towed away for scrap.

Back at the Petwood Hotel, Knilans stood in the courtyard, stripped naked and burned all his louse-infested clothes, before going inside to take a welcome bath and find a set of clean clothes. He had flown twenty ops with 619 Squadron before joining the Dambusters, and had now completed another thirty ops with 617. He had won the DSO and DFC, and his own countrymen had awarded him their DFC twice and five Air Medals. Willie Tait had also put him up for a Bar to his DSO, but it was vetoed by the Station Commander, probably because of the time Knilans had 'buzzed' Petwood Hall while the Group Captain was having his afternoon tea! Knilans had survived fifty ops and brought six of his original seven-man crew through the war alive, but he was the only man of the thirty-two trainees on his pilots' course to survive the war. After a period of rest, he volunteered to fly American 'Black Widow' night-fighters in the Pacific, but the war ended before he could do so.[37]

Despite the suggestion from aircrew and surveillance flights that the *Tirpitz* was relatively unscathed after the first raid by 617 Squadron, unknown to them, 'one bomb had knocked the guts out of it and finished it as a fighting unit for ever.'[38] One of the Tallboys, almost certainly Tait's, had detonated close enough to *Tirpitz*'s keel to blow a 1,500-square-foot hole in the starboard bow, so large, according to Klaus Rohwedder, that 'a small motorboat can turn inside'.[39] Torrents of seawater flowed into the ship and, together with internal strains and damage from the near-misses, rendered the battleship unseaworthy. German engineers estimated repairs would take nine months, even if no further damage was sustained, but a committee chaired by Admiral Dönitz concluded that the *Tirpitz* was already beyond repair.[40]

Rohwedder also testified to the human cost of the attacks on the *Tirpitz*, his voice cracking as he recalls the horrors:

The military hospital also had a bomb hit and all the doctors were dead. There were just a few paramedics left alive who cared for the wounded. There was no light and the men moaned and howled. They cried for their mothers. That still sticks with me. No one cried for their father. They cried for their mother and ... and then ... I couldn't stand it any longer ... That day we had one hundred and thirty-two dead people and about three hundred wounded.

The 'butcher's bill' from the repeated attacks on the *Tirpitz* was something that Rohwedder now knew only too well. 'It is a grue-some affair,' he says. 'The hero's death was always portrayed as a splendid deed, but how dirty death is and how badly people had to die, I learned on this day.'[41]

A month after the raid, a surveillance flight by a Mosquito over Altenfjord reported that the *Tirpitz* had disappeared from its mooring, causing fresh alarms throughout the British high com-mand and a frantic search. Within a few days an aircraft carrying out a recce flight over Norway sighted the battleship off the island of Haakøy in Tromsø Fjord and sent an urgent signal: '*Tirpitz* in Tromsø this afternoon. Big hole in forward deck.'

The gaping hole in the bow – the product of Willie Tait's direct hit with a Tallboy – had been hastily patched with steel plates and the crippled ship had then been towed at a painfully slow 6 or 7 knots round the coast from Altenfjord. North Sea storms and attacks from British aircraft would have made any attempt to return the ship to Germany for repairs much too risky, and instead it was towed into a sheltered, shallow-water anchorage in Tromsø Fjord. It was now a safer distance from the Russian forces steadily advancing from the east, and though it could not put to sea, its guns could still be used to repel a feared Allied landing in northern Norway.

However, still unaware of the full extent of the damage to the ship, British commanders continued to regard it as a potent threat to the Arctic convoys, and its new location, 120 miles closer to Britain than Altenfjord, had brought it just within the return range of Lancasters flying from RAF Lossiemouth in northern Scotland.

A second attack, Operation Obviate, was immediately authorised whenever weather conditions permitted – a severe limitation at Tromsø, which was under semi-permanent cloud cover in the prevailing westerlies and was certain to be cloud-free only in an easterly wind.

CHAPTER 12

'What Have You Been Doing Today?'

The crews who returned from Yagodnik were soon receiving sad news. Wing Commander Guy Gibson, VC, DSO and Bar, DFC and Bar, the leader of the squadron on the legendary Dams raid that had made it immortal, was killed on 19 September 1944 during a raid on Mönchengladbach. A man born to fly, with over 170 missions under his belt, Gibson had been unable to resist the urge to fly on operations once more.

He had been posted to 54 Base as operations officer (a non-operational role), but took advantage of his position to place himself on the battle order that night as 'Master Bomber', controlling the whole attack from over the target area – a role for which he had no formal training. There was considerable difficulty marking the target, leading many aircraft to spend longer above it than was desirable. Gibson himself may have circled the target once more after the other aircraft had turned for home. Whether he was hit by flak or fire from a night-fighter, suffered a mechanical failure, made an error because of his unfamiliarity with the Mosquito he was flying, or simply ran out of fuel is unknown, but his aircraft crashed at Steenbergen in the Netherlands at 10.30 that evening, and Gibson was killed instantly.

The Dutch mayor of the town arranged a proper funeral for Gibson and his navigator, Squadron Leader Jim Warwick, DFC, who, as Station Navigation Officer, should not have been flying combat ops either. Winston Churchill paid full tribute to Gibson, 'the glorious Dam-buster', in a letter written to his sister: 'We have lost in this officer one of the most splendid of all our fighting men. His name will not be forgotten; it will be for ever enshrined in the most wonderful records of our country.'[1]

Because of the propaganda value the Nazis might have made from it, Gibson's death was not officially announced until 5 January 1945, but the RAF grapevine had spread the news to 617 Squadron almost at once, and those who had flown with him, whether or not they liked him, were united in respect for the skill, courage and leadership he had shown when commanding the squadron to its most famous achievement.

Frank Tilley

Before they could return to the task of finishing off the *Tirpitz* in that autumn of 1944, the men of 617 Squadron were first tasked with making two other major raids, one a return to the scene of the squadron's worst disaster: the Dortmund–Ems Canal. That op was set for 23 September 1944, less than a week after they had returned from the first attack on the *Tirpitz*. Squadron Leader Wyness had lost two of his regular crew in the crash in the Norwegian mountains that killed all of Flying Officer Frank Levy's crew, and one of the replacements was a twenty-one-year-old flight engineer, Frank Tilley, who had never flown on a

night operation before and was told only that afternoon that
he would be flying with Wyness. 'It was my first night trip and
everything seemed more forbidding,' he says. 'I felt very apprehensive about it. I preferred day to night; I felt much happier if
I could see out!'[2]

Tilley, a Londoner with a roguish smile and a chirpy manner,
had only joined 617 the previous month. He had worked at an
engineering firm on London's Whitechapel Road after leaving
school, and his first direct experience of the war had come in mid-
1940 when he was seventeen. The Germans bombed the docks,
and sheltering in the basement, he saw the bombers flying over,
heard the explosions, saw the flames coming up and the raging
fires. 'It was quite scary,' he says. 'We just hoped they wouldn't
bomb us! Eventually they did bomb the factory and that was my
job there finished!'

Tilley didn't think that much about the war at first, even
though several of his friends enlisted. 'Some really wanted to get
stuck in,' he says, 'but I didn't feel that way in the early days, I
don't think I was very mature and felt quite young even at seventeen. But eventually I succumbed to the tremendous propaganda
– the RAF, especially the bombers, were really glamorised – and
I thought I needed to be part of it all.' He was working for
a company manufacturing artificial limbs at the time – always
a growth business in wartime, but 'a very boring job', he says. It
was a reserved occupation and he could not enlist in the normal
way, but he discovered that if he volunteered for RAF aircrew
then he could be released to fight. 'I had no idea about the huge
aircrew losses,' Tilley says, 'though logic should have made me
ask, "Why is it so easy to volunteer as bomber aircrew?" and
wonder what had happened to all the others from the previous
years, but of course you didn't think about that back then!'

Tilley trained as a flight engineer, and at the end of his training
he assembled in a hangar with the other new flight engineers,
facing a line of officers, who began calling out, 'We need
ten people for Sunderlands' – the flying boats mainly used on
anti-U-boat patrols by Coastal Command. 'There was a mad

rush,' Tilley says. 'Clearly, the others had realised the dangers of flying bombers!'

The officers then asked for volunteers for other aircraft, and Tilley found it 'rather thought-provoking' that the largest numbers of volunteers were required for the Lancasters, Halifaxes and Stirlings used by Bomber Command. He volunteered for Lancasters mainly because he liked the look of the aircraft, and was accepted at once. 'I suppose I must have realised by then that it was a more dangerous option,' he says, 'but I think I wanted to be at the forefront of it all.'

He joined New Zealander Arthur 'Joppy' Joplin's crew simply because Joplin walked straight up to him and asked him to. 'I've no idea why,' Tilley says, 'but that was it. The relationship with the crew was one hundred per cent important. If you didn't get on, then you were in serious trouble, but in fact Joppy was a wonderful pilot and a wonderful captain, and got on with everybody. So I was now part of a crew which would live or die together.'

He became very good friends with the crew's navigator, Basil Fish. 'He had his dad's car,' Tilley says, 'a dark blue, two-door Ford Eight, and we used to sneak out to the squadron in the middle of the night and pinch petrol for it so we could get around! We all used to cram into it for a night out in Boston. Before then I didn't drink or anything, I was quite innocent really, but Basil introduced me to Black and Tan and I never looked back!'

'Every boy in the world wanted to fly,' remembers Basil Fish,[3] not least because being aircrew 'didn't hurt when you went to a dance on a Saturday night. You were hoping some girl would ask, "What have you been doing today?" so you could say, "Oh, I've been dicing with death up in the heavens"!'

'We were popular with the girls,' Frank Tilley says:

most of us aircrew were. One particular girl took a real shine to Joppy – he was the good-looking one – and being a New Zealander made him an extra special catch! We used to go to The Gliderdrome in Boston, a skating rink turned into a dance hall. They would have a band and it ended up as our regular place to go after the pub. Looking back now, it's curious to think of us dancing and socialising then heading off next day to drop bombs, destroy buildings, probably kill people. But we didn't think like that then – it just didn't cross our minds.

As they trained on the Lancaster, the stories from front-line crews about going on ops, flak and being shot up by fighters made them realise that 'we were heading towards hell when we started ops,' Tilley says, 'but we didn't want to let anyone down, and we certainly didn't talk about fears or worries.'

Their skipper, Arthur Joplin, who had joined the war effort 'to be part of it all, to experience the excitement', agrees that any fears simply didn't seem to surface. 'I don't think we ever really talked as a crew about the dangers of flying in Bomber Command,' he says from his home in Auckland. 'And I certainly didn't think too much about it all – we just lived life as it came along. And to the full! I was more concerned with ensuring I managed each op correctly – there was so much to do and to learn, everything was very new and you really had to concentrate to ensure you kept on top of everything.'

With Joplin and the rest of his crew, Tilley was posted to 617 Squadron on 15 August 1944, but says that at the time '617 was just another number to me. They weren't regarded as a heroic squadron or anything, just a "special duties" squadron. I'd heard about the Dams raid, but it wasn't big folklore in the RAF, I didn't even associate 617 with it, and in fact we didn't even call it "the Dams raid", it was just another op, and the term "Dambusters" didn't exist until well after the war when the film came out.' Tilley's memory may be at fault on that point, for the *Daily Mirror*'s headline on 27 May 1943 was 'Dam-buster Gibson to get VC'. But Joplin remembers events differently: 'I didn't have a clue about 617 Squadron back then! Back in NZ training I don't even remember hearing about the Dams raid.' Tilley's crewmate, Loftus 'Lofty' Hebbard, had a rather more jaundiced view of their new unit. After having a scout around, he commented to his crew, 'It looks like we have been posted to that suicide squadron.'[4]

Frank Tilley was about to have a worrying introduction to 617 ops, and his apprehension about the Dortmund–Ems op would have increased had he known what had happened on the previous raid on the canal in 1943. Perhaps fortunately, he had 'no idea at all that just a year before, five out of eight aircraft had been shot down there, but I shall remember the trip as long as I live; it was dreadful.' Because the canal was such a 'hot', well-defended target, 617's raid was timed so that they flew out alongside a large contingent of Main Force bombers that were targeting the city of Dortmund. The sheer numbers of other aircraft gave 617 Squadron some cover from prowling night-fighters, but even so, Wyness's Lancaster was attacked by fighters several times, both on the approach to the target and after turning for home.

Unlike Tilley, the other members of the crew were very experienced and responded to the attacks with well-drilled coolness. Tilley had no idea that they were even under threat until he suddenly heard the rear gunner's urgent call, 'Corkscrew port skipper!' An instant later, as the Lancaster began its vicious, twisting dive, Tilley saw tracer flashing across the top of the aircraft. 'I could see it clear as day. It was a constant stream of green tracer and those bullets spelt death. I can still see it today,' he says, holding out his hands to show the tremor that has started in them as he recalls the incident seventy years later from the comfort of his study ain his home in leafy Hertfordshire.

Wyness had to corkscrew several times to shake off fighters. 'I had no part to play in it,' Tilley says. 'I mean, he'd got the control column, I hadn't. It was up to the gunners and the pilot. I just clung on because when you dive you almost float off the floor, but then, when he rolls and starts climbing, you're crushed back down.'

The attack lasted about two or three minutes but 'it felt like a lifetime'. Tilley had experienced the realities of battle previously and says 'there was nothing to do but monitor the instruments and stare at the tracer searing through the surrounding darkness'. 'I was transfixed by it,' he says, 'and by seeing other Lancasters going down in flames.'

It was a shocking realisation. In that moment I realised what the war,
Bomber Command, the flying, all meant. It was the first time I'd seen
anything like that. In the total blackness, you'd suddenly see a ball of flame
which got bigger and bigger, and then just descended gently downwards to
the earth. Knowing that it was a Lancaster with at least seven people dying
in front of your eyes was truly awful – it's an image that I carry with me to
this day – and it went on and on, perhaps eleven times altogether. At the
time I suppose I just thought, 'Thank God it's not me,' but I find myself
thinking about it more now, and understanding much more what it meant.
Some years after the war, a friend described me and the other bomber air-
crews as 'just cannon fodder'. I took exception to that at the time, but I
suppose he was right. Just like the trenches in the First World War, you
knew that a lot of you wouldn't be coming back.

Only one of the Lancasters shot down that night was from 617
Squadron, with the loss of three lives, the remainder being Main
Force aircraft, and this time, unlike their previous raid on the
Dortmund–Ems Canal, their bombing was accurate, destroying
part of the embankment. Ten miles of the canal were completely
drained of water, leaving twenty-three barges high and dry and
flooding the surrounding countryside. A workforce of 4,000 men
laboured for weeks to repair the breaches that had appeared along
2 miles of its length, but the canal remained unusable for the
remainder of the war.[5] The one major blot on 617 Squadron's
record had now been erased.

Frank Tilley was never more glad to see the runway lights at
Woodhall Spa than he was that night, but if he was expecting
some acknowledgement of the baptism of fire by flak and night-
fighter he had been through, Squadron Leader Wyness was to
disappoint him. After they'd landed, Wyness merely said, 'Oh,
well done, Engineer. What was our fuel consumption?'

'Well, heck, I told him,' Tilley says with a rueful smile, 'and
he just said, "Thank you very much," and that was that – my
first experience of a night operation.'

★   ★   ★

Despite the failure of Operation Market Garden – the heroic but doomed attempt by airborne troops to seize the bridge at Arnhem – Allied successes had continued during the early autumn. Brest, Boulogne and Calais all fell to the Allies by the end of September 1944, while on the Eastern Front, Soviet forces advanced through Yugoslavia and Hungary. In mid-September, Sir Arthur Harris had been formally released from his orders to support the invasion and Bomber Command reverted to Air Ministry control as attacks by Main Force resumed on German industrial targets. In October, 17,000 sorties were flown, 13,000 of which were to targets in Germany during which 50,000 tons of bombs were dropped. On 14 and 15 October, Bomber Command attacked Duisburg with thousand-bomber raids that dropped the same weight of bombs in less than forty-eight hours as the Germans had dropped on London throughout the war. The Duisburg official archive details local reports of the time saying, 'Very serious property damage. A large number of people are buried. All mines and coke ovens lay silent.'[6]

Meanwhile, 617 Squadron continued its attacks into the heart of Nazi Germany by day and night, and although the war may have been approaching its endgame, the dangers were still all too real.

⋆   ⋆   ⋆

'Heads!' Bruce Hosie – a New Zealander who was still only twenty-one but already the veteran of seventy-two ops, forty-seven of them with 617 Squadron – was spinning a coin with another wireless operator to decide who would fill a vacant seat in Drew Wyness's Lancaster. It was 7 October 1944 and Wyness's regular w/op had an ear infection and couldn't fly, so Hosie and the other w/op were competing for the role. The other man called correctly, earning himself the day off and committing Hosie to a raid on the Kembs barrage, something of a poisoned chalice, and an even more difficult and heavily defended target than the Dortmund–Ems Canal.

Born in the farming town of Mania, Taranaki, Hosie was barely eighteen when he left the family home to join up. His uncles and father were veterans of the First World War, and with the heavy losses suffered in the trenches, there was a certain amount of dread within the family as their only son departed. But, despite the dangers of continuous ops over Germany, Bruce proved to have a lucky streak – he was in the sick-bay when his original crew on 617 Squadron was shot down over Munich. Tall and blond, Hosie enjoyed the social life on the squadron, and his diary reveals regular trips to the watering holes of Lincoln both with his crew and with a number of girlfriends. But he also missed his family and home, and there are countless entries detailing the letters he wrote to his loved ones, and the precious notes received in return. On the day he lost the toss, he had already completed the equivalent of almost three Bomber Command tours and could easily have opted to be repatriated to his homeland. But, like so many others before him, he was dedicated to the fight and, in the aftermath of D-Day, was determined to see the war won.[7]

The Kembs barrage, close to the Swiss–German border a few miles north of Basel, was part a hydroelectric scheme, but its huge sluice gates also controlled the flow of the Rhine, maintaining it as a navigable waterway. As US commanders planned the crossing of the Rhine, they were haunted by the fear that the Nazis would deliberately destroy the barrage, creating a cataclysmic flood that would sweep away the American assault troops and pontoon bridges as they crossed the Rhine further north. Thirteen Lancasters from 617 Squadron were therefore tasked with making a pre-emptive strike, aiming to destroy the barrage and release the pent-up waters long before the US troops reached the banks of the river. The ensuing devastating flooding would instead create chaos among German units moving to defend the Rhine. The final briefing for the op contained one significant warning: a heavily defended powerhouse lay just to the north-west of the target and crews were warned to avoid its flak batteries at all costs.

It was felt that a daylight raid would give the maximum chance of success, and three squadrons of Mustangs were detailed to provide fighter cover. The Lancasters were divided into two groups. Seven Lancasters, led by Squadron Leader Gerry Fawke, would bomb from 7,000 feet, with the aim of distracting the air defences. Meanwhile, the remaining six, led by Willie Tait, would sweep in at low level in three pairs to attack the barrage. They were using delay-fused Tallboys both to stop smoke and debris from the previous bombs obscuring the target, and to avoid being hit by the blast or shrapnel from their own weapons.

However, the plan's chances of success rested on some dangerous assumptions, not least that the German air defences would not be alerted by reports of a flight of Lancaster bombers turning down the Rhine at the Swiss border, even though the Kembs barrage was the first and by far the most significant target in their path. Even more worrying, all the aircrews were very well aware that Lancasters making a low-level daylight attack against a well-defended target would be extremely vulnerable to flak and ground fire, and most had serious misgivings. They would be approaching the target at around 600 feet, an altitude that one pilot described as 'madness – ideal for an [enemy] machine gunner'.[8]

The day before the op, Tony Iveson had lunch with Drew Wyness, Bruce Hosie and Colonel Chris Melville, 'the Commanding Officer of the King's Own Scottish Borderer's – they had been training in Scotland for mountain warfare, and were a very tough bunch! He drove us back in his jeep, and dropped Drew and me at our aircraft. As he left we arranged to see him for drinks after the op next day. We all thought the war was going well and were looking forward to a conclusion soon.'[9]

The Mustangs linked up with the Lancasters over Dungeness after an afternoon take-off on 7 October 1944, and the flight to the target, in fine, sunny conditions, was peaceful enough. Flight Lieutenant Phil Martin had joined 617 from 61 Squadron, where he had been awarded his first DFC for completing thirty missions, and his navigator recalled 'as we ambled along, quite low down, seeing a farmer and his cows, each with a bell around its neck, crossing a bridge. I have never seen cows with bells on, only in books. Quite a peaceful scene, soon to be forgotten in the next few minutes!'

The weather was touch and go nearer the target, but the barrage itself was clear of cloud below bombing height and visibility was good. The low force did their planned dog-leg as Squadron Leader Gerry Fawke's group climbed to bombing height, meeting only minor flak. From the target 3 miles ahead, Willie Tait could see heavy fire being directed at Fawke's formation, followed by a series of splashes as their Tallboys struck the river.

The Mustangs dived out of the sun and began shooting up the flak batteries in an attempt to divert their fire from the low-level bombers, but the fire from the air defences was much heavier than expected and, though none of the attackers had been shot down so far, their luck could not hold indefinitely. Exactly on H-hour, Tait's Lancaster dived out of the 1,500-foot cloud base and levelled off at 500 feet. He could see explosions and eruptions like waterspouts as the high-level force's Tallboys detonated. He thought for a moment that the flak-gunners had not seen him, but then lines of white tracer seared upwards from the east bank of the river. All traces of the bomb-bursts from the high-level force had now disappeared, but there appeared to be no damage to the target, with all of the sluices still closed and no cascade of water.

Tait's bomb-aimer, Walter 'Danny' Daniels, squinted into his bombsight and the aircraft jumped as the bomb dropped away. He watched it enter the water towards the left-hand end of the barrage and sink like a stone as planned. Tait slammed the throttles forward and heard his rear gunner open fire as they passed over the barrage.

As they swept away on a starboard curve, one of Tait's crew thought he 'saw the kite immediately behind us burst into flames, go into an uncontrolled dive and explode on the ground'. In fact, Drew Wyness, following his leader in, found flak flying all around him, 'the most frightening barrage of light and medium flak that we have ever seen during the war'.[10] Flying at low level, in broad daylight, the Lancaster was an easy target as, ignoring the efforts of the Mustangs to distract them, German gunners poured shells and bullets into it.

Hit repeatedly and streaming smoke, Wyness's Lancaster burst into flames before reaching its bombing point, and though the Tallboy was released, the aircraft was now almost uncontrollable. It hit the high-tension power cables leading from the barrage, causing huge blue flashes as the cables shorted. Then, shortly after hitting the water near the bank of the Rhine, the aircraft exploded, leading Willie Tait's crew to believe that every man on board had died at once.

Kembs barrage under attack

The remaining raiders also met murderous fire as they targeted the Kembs barrage. John Cockshott's Lancaster hit the slipstream of the previous aircraft and his Tallboy fell wide. Australian pilot Jack 'Slapsie' Sayers made two runs, but as he opened his bomb doors on the second one, either his Tallboy had been inadequately secured in the bomb-bay, or an electrical fault had caused it to release prematurely, because it tumbled straight out of the bomb doors as they were opened, buckling one of them. The bomb fell to earth and landed in soft ground without detonating. It was retrieved intact by the Germans, giving them another chance to examine the construction of this new British super-weapon.

Sayers had earned his unusual nickname after an incident when, seeing a drunken man and his friend beating up a woman, he intervened and knocked out both of them. He then beat a hasty retreat after the woman took a fancy to him. His mates christened him Slapsie after a ferocious puncher, the light-heavyweight boxing champion 'Slapsie Maxie' Rosenbloom.[11]

The next pilots to attack the target, Christopher 'Kit' Howard and Phil Martin, misjudged their approaches and banked around to make another run. Howard's decision to go round again, ignoring Willie Tait's instruction 'Kit! Abandon, abandon!',[12] proved a fatal error. The German gunners had watched the heavy-laden Lancasters slowly circling and unleashed a blizzard of fire as Howard began his run. His starboard fuel tanks were ruptured and the plane burst into flame. Still raked by cannon fire, he veered away to starboard and moments later the aircraft exploded as it hit the ground near the village of Efringen-Kirchen, just inside Germany. Howard, whose elder brother had already been killed in action, was from one of Britain's oldest aristocratic families and had been heir to the estate of Castle Howard in Yorkshire.[13]

Flying 400 yards behind him, Phil Martin's gunner saw Howard's aircraft explode. 'This made me very angry,' he said. 'I opened up my guns and gave them everything I had. I can remember seeing the tracer bullets going into the defences with men running about.'[14] As their bomb-aimer called 'Bomb gone!' the aircraft shuddered when it too was hit by flak, damaging the rudders. Had a Mustang not flown directly underneath him, engaging the anti-aircraft guns and distracting attention from the Lancaster for a few crucial seconds, Martin too might well have been shot down. As it was, he escaped the guns, but his bomb overshot by 50 yards. Struggling to control his turn, Martin then found himself flying through a balloon barrage. His mid-upper gunner asked if he could have a go at them, but was immediately told, 'No! They're Swiss!'

They missed the balloons, but Martin was getting no response from the rudder and their chances of returning safely to England did not look bright until the flight engineer made his way to the back, wrapped the severed rudder wire around the handle of his fire axe and used it as a rudimentary rudder control. When they landed safely back in Woodhall Spa, they counted 106 flak holes in their aircraft.

Tait had climbed to 8,000 feet, but then, as was his habit, he decided to go around once more to see if any of the others had scored hits. While doing this, forgetting his own warning to his men, he flew straight over the powerhouse, which was defended by two four-gun batteries of light flak. 'In a few seconds they were on us and followed us all over the sky until we went out of range.' Despite having one tyre shot away, a 40mm flak shell lodged in his wing and his main fuel tank 'leaking like a colander', Tait made a safe landing back at Woodhall Spa.[15]

As the battered Lancasters were flying home, a Mosquito from 627 Squadron made two runs over the target to photograph the results as the delay-fused Tallboys detonated. Several exploded harmlessly, but on their first run they saw one bomb burst some 200 yards south of the west end of the barrage, soon followed by another 'which appeared to blow out the westerly span. Water started to pour through the gap and there were ripples extending 200–250 yards upstream.'

Tait's bomb had come to rest against the left-hand sluice gate and when it and Jimmy Castagnola's bombs exploded, they demolished the gate and the iron superstructure above the first and second pillars on the barrage's west side, unleashing a wall of water that emptied the dam, flooding the lower Rhine valley and causing the water upstream to fall dramatically. The Swiss newspaper *National-Zeitung* reported: 'The breaching of the Kembs Dam has lowered the water level in the Rhine basin at Basle, necessitating the transfer of boats from the first basin to the second. At 2100 hours the level of the Rhine fell by three to three and a half metres. Below Kembs the water released is estimated at millions of cubic metres and has apparently caused flooding everywhere, for the German authorities have given the water alarm.'

★ ★ ★

The repercussions of the attack were far from over. On 9 October Willie Tait, still presuming that Drew Wyness's crew had all perished when he saw their Lancaster explode, sent a letter to Bruce Hosie's mother in New Zealand expressing his deepest sympathy for her loss. 'I know you would prefer me to be frank with you,' he wrote. 'I think that all the crew must have been killed instantly.'[16] In fact and miraculously, Hosie and his crew all survived the impact after Wyness managed to ditch his Lancaster in the Rhine, coming to rest in shallow water near the western bank, close to the Franco-German border town of Chalampe.

Bruce Hosie

Two of the crew ran along the wing, jumped on to the riverbank and made off into the woods flanking the river, while the other five inflated their dinghy and began paddling downstream. The burning aircraft they had left exploded after they were clear.

When they saw the Germans launching a boat to intercept them, one of the five men dived overboard, swam to the bank and made for the woods, but the other four remained in the dinghy and were captured. They were taken to the nearby town of Rheinweiler, where the local Nazi chief, Kreisleiter Hugo Grüner, took charge of them. Unknown to the airmen, since August 1943, Nazi leaders, furious at the impact of the Allied bombing campaign against Germany, had been giving tacit encouragement, and sometimes direct orders, for reprisals against captured British and American *Terrorflieger*, 'terror flyers' – 'lynch law should be the rule'. Even without such official directives, aircrew baling out were often at risk: in the course of the war, 350 aircrew were butchered by civilians furious at the bombing they had endured.

The Gauleiter of the Baden and Alsace region in which they had crash-landed had then issued a specific order that Allied airmen captured in the region were to be killed. His subordinate, Grüner, did not disobey. Two at a time, he took Wyness and his crewmates to a deserted stretch of the Rhine a little further downstream and executed them.[17]

Captured after the war, he confessed that 'I murdered them by firing a machine-gun salvo at each of them in the back, after which each airman was dragged by the feet and thrown into the Rhine.' Their bodies were later discovered miles downstream. Among them were those of Wyness himself and the New Zealander Bruce Hosie, who had joined the crew only at the last moment. Hosie, the man who had spun a coin with another wireless operator to see who would fill the seat in Wyness's aircraft, had lost the toss and, with it, his life. The three crew members who had escaped into the woods did not fare any better. They were never seen again and it is believed that they were also summarily executed and buried in unmarked graves. Grüner escaped from American custody in 1947 and though convicted *in absentia* of murder, he was never recaptured.[18] Hosie's great-nephew says the family 'never really got over his death ... ANZAC Day is always a pretty emotional day around home for lots of reasons, Bruce being one.'[19]

★   ★   ★

Tait's bomb-aimer was awarded 'an immediate DFC' for destroying the barrage and Phil Martin an immediate Bar to his DFC.

Four other men who had taken part in the raid, including two more of Willie Tait's crew, received the DFC, but there was no medal for their deceased comrade, Kit Howard, who had flown sixty ops without being awarded any decoration. Some of his comrades tried to get him a posthumous medal, but in the event, no award was ever made.

Tony Iveson felt the loss of Drew Wyness particularly keenly. 'We had become mates. He was tall, fair-haired, full of beans and probably one of the best-looking young men I ever knew. His great interest was women! He would chat up almost any girl – I have often wondered if there was something inside him that said, "Get on with this, enjoy it, because you are not going to be around long."' Iveson had the sad task of telling their mutual friend, the army officer Chris Melville, whom they'd lunched with the day before, that Drew would no longer be joining them for their prearranged drinks. 'When I told him, this rugged tough guy, who'd played rugby for Scotland, just sat there with tears running down his cheeks.'

Looking back on the raid years later, Willie Tait felt that the whole plan had been 'too complicated. The high-level force did not divert the flak, it alerted them. Likewise the fighters. It would have been better to have two or four aeroplanes on the low-level and nothing else. We would probably have slipped in without anyone being alerted.'[20]

★   ★   ★

The Kembs barrage operation never attained the fame of the Dams raid, but it was no less difficult and dangerous. The threat to advancing American forces had now been nullified, but once more, 617 Squadron's men had paid a heavy price for their success, losing fifteen of their comrades.

CHAPTER 13

Back to the *Tirpitz*

The Tirpitz *in a Norwegian fjord, 1944*

The attention of 617 Squadron's crews now returned to the *Tirpitz*, still lurking in the icy waters of Tromsø Fjord. In preparation for the attack at maximum range, ground crews worked frantically to replace all the Lancasters' engines with newer and more powerful Rolls-Royce Merlin 24s, paddle-blade propellers and long-range fuel tanks. In the place where the rest bed and oxygen bottles normally stood, ground crews fitted a Wellington fuel tank with a Mosquito fuel tank secured on top. Fuelling these additional tanks was so hazardous that the ground crews were made to wear plimsolls rather than boots, because 'one spark could have blown up the aircraft and probably the ones to either side of it as well'.[1] Even though all the fuel joints had been tightened and there were no visible signs of any leaks, there was still a pungent enough smell of 100-octane fuel inside the aircraft to make several crewmen feel queasy. It was also 'like flying a giant fuel tank', which flight engineer Frank Tilley, with commendable understatement, described as 'a bit worrying'.[2]

Although 'Joppy' Joplin's crew, including Tilley, had flown their first op, a raid on Brest, on 27 August 1944, their inexperience had obliged them to sit out the first attack on the *Tirpitz*. Now they were to join their comrades in the second attack. Every scrap of excess weight was removed from their aircraft, including the armour plating protecting the pilot's seat. The mid-upper gun turret was completely removed, the guns were taken out of the front turret, and the ammunition for the rear guns – the Lancaster's only remaining defensive armament – was reduced to just 500 rounds.

On 28 October 1944, after receiving a favourable weather forecast, twenty 617 Squadron aircraft, once more augmented by a similar force of Lancasters from 9 Squadron, took off. By now, bomb-aimer Keith 'Aspro' Astbury had finished his tour of duty with 617 Squadron and been posted to the Air Ministry in London, but he found his duties so excruciatingly boring that when a fellow Australian, Arthur Kell, told him his bomb-aimer had fallen ill and could not make the *Tirpitz* op, Aspro wrote himself a forty-eight-hour pass and caught the next train north. By the time the Station Commander at Woodhall Spa, Group Captain 'Monty' Philpott, found out that Aspro had no authority even to be there, let alone to be going on an op, he was already airborne.

Half of 617's Lancasters landed at Lossiemouth, the remainder at Milltown, while 9 Squadron used Kinloss. After refuelling and final briefing, they planned to take off again at 1 a.m. the following morning, 29 October, but not before a few of the more enterprising souls had managed to scout around the base. Pilot Lawrence 'Benny' Goodman met 'a very pretty young WAAF a few hours before take-off', he recalls, smiling at the memory. 'It was very dark, and as I went around a hangar I literally bumped into her and nearly knocked her over. We started chatting and I asked her if she'd like to meet up again. I explained I wouldn't be free that night as I had to go flying, but perhaps we could meet up the next day? She said she'd love to, so I went off with a skip in my step to prepare for the op.'

Benny clambered aboard his aircraft knowing the length of the return flight would put the strongest bladder to the test – luckily, the WAAFs from the Mess at Lossiemouth had thoughtfully provided each crewman with an empty milk bottle to pee in. He waited to take off on another miserable night – 'it was pitch black with a low cloud base and rain hammering down'. As Tony Iveson turned onto the runway and wound up his engines for take-off, such was the massive load they were carrying – around 68,000 pounds – that to get airborne he had to push the throttles 'through the gate' – right to the maximum – which 'could ruin an engine quickly, but we had to do it because of the load'. However, as they rumbled down the runway, one of his port engines failed to reach full power, and only quick action by his flight engineer, who pulled back on a starboard throttle, balanced the aircraft as it took off.

Struggling to get airborne, only inches off the ground, they were heading straight for Benny Goodman's Lancaster, lined up on the peri-track awaiting his turn to take off. The two aircraft, brimful of fuel and each carrying a 12,000-pound Tallboy bomb, were now only seconds from a devastating collision. Goodman 'had my head inside the cockpit doing checks, and suddenly Jock, my flight engineer, nudged me hard in the ribs and said, "Look!" I looked up and saw a bloody great undercarriage heading straight for us. It looked like it was coming straight through the cockpit. I confess I ducked!' A split-second later, Iveson's Lancaster roared overhead, missing the other aircraft by inches. Soon after, Goodman himself 'lined up and took off, nerves over, forget it!'

Part of the flight was over neutral Sweden. There was no blackout there and Frank Tilley found looking at all the lights of the towns 'amazing, as I'd not seen that before. A little flak came up but I had the impression they weren't really trying to hit us. Just a gesture really.' As they flew on, he found the mountains 'a beautiful sight – snow-capped, serene and peaceful. Quite a contrast between the scene unfolding before us and what we were actually there to do: destroy a ship and kill people.'[3]

At these latitudes in late October, the sun remained low in the sky all day and the dawn produced only a gradual strengthening of the light. One Australian rear gunner was watching as 'the daylight struggled through the darkness, making an eerie break to the day,' but as he peered into the lightening sky to the east, he jolted upright. Just forward of the starboard beam and flying on an exact reciprocal course was a German transport aircraft:

a lumbering old Ju 52, tiredly staggering its way southward towards Germany, those within probably eagerly anticipating the coming leave and certainly not expecting to see enemy aircraft heading towards the North Pole! We were not spotted. If we had been, none of us would have come back to Lossiemouth that night, the fighters would have made sure of that. That is why I did not fire at a target which looked close enough to reach out and touch![4]

The sun rose above the horizon, 'tinting all the mountain tops pink, just like a wedding cake',[5] but they could see banks of cloud ahead of them. The weather over Tromsø had been predicted to be fine and clear, but as the Lancasters approached they found that cloud cover streaming in from the sea was obscuring the target. Nonetheless, they pressed home the attack through a blizzard of anti-aircraft fire. As usual, wireless operator Sydney Grimes was standing up in the astrodome of his aircraft, watching the action:

We'd come a long way to hit the ship and I wanted to see what she looked like! I could see the flak coming up and exploding. The heavy fifteen-inch guns could reach us quite far out, so it was like flying into the side of a house. It was a huge explosion but luckily they hadn't got the height quite right, so they were exploding beneath us. It was quite strange watching these massive blasts; they were almost beautiful, but if one of those had got near, we'd have been dead.

Grimes came from Great Wakering, near Southend. His father was a Thames bargeman who during the First World War had ferried supplies of ammunition from Woolwich Arsenal across the Channel to Saint-Valéry, near Dieppe. When war broke out, Sydney, then seventeen, was working as a clerk in E. K. Cole's radio factory – it later adopted the brand name EKCO and made the radios that Sydney now used as a wireless operator in the RAF. During the Battle of Britain, he and his friends had watched dogfights taking place over the Thames estuary. 'We saw aircraft shot down and then cycled as close as possible to the site, to see what it was like.' From the train between Southend and Fenchurch Street in London, he had also had a close-up view of the effects of the Blitz: 'flattened buildings, debris still smouldering – it really was the most horrific sight. It made me realise I needed to be part of the war, to take the war back to the enemy. I joined up because everyone else was; I wanted to get in before it was over. I didn't want to live the rest of my life having missed out when all my friends were part of it.' He also had another motivation: a close friend of his had joined the Navy and had drowned when his ship was sunk in the Mediterranean. 'I thought to myself, I will have to meet his family some time, see them and talk to them, and if I haven't "done my bit", I will be very embarrassed.'[6]

Grimes couldn't swim, which ruled out the Royal Navy, and he didn't want 'anything to do with the Army in case there was trench warfare', so that left the RAF. 'With one eye on the end of the war and what I might do,' he decided to learn a new skill and get a trade by becoming a wireless operator. After training, he joined 106 Squadron, and such was the carnage among Bomber Command's Main Force aircrew that Sydney's crew became only the second from his squadron ever to survive their thirty-op tour of duty. 'You'd come back from a meal or briefing,' Sydney says, shaking his head at the memory:

and the Committee of Adjustment [the men tasked with clearing out the lockers and personal effects of men killed or missing in action] would be clearing away belongings. One day we woke up and they were clearing the personal items away: photos, clothes, everything, just going in a big sack. Another crew gone. It was quite a stark image. But you couldn't let yourself

think about that – it would never happen to you! Of course we all said that,
didn't we? And it happened to half of Bomber Command.

To dwell on the losses was to invite uncomfortable reflections on their own slim chances of survival, and thoughts of the dead were usually set aside as swiftly as their possessions were whisked away by the Committee of Adjustment. That was always done within hours, partly on the principle of 'out of sight, out of mind', but also because, in the brutal pragmatism of combat squadrons, the bed and locker space would be needed at once for the dead man's replacement.

Mac Hamilton illustrated the necessarily harsh realities of life on a squadron where some men would not return from almost every op. After one night raid, Hamilton fell into bed about six in the morning and had just got to sleep when he heard:

> *this clump, clump, clump, and thought, What the hell are they making all*
> *that noise for? I got up and there were these two corporals with red bands*
> *round their hats. I said, 'What the hell is going on?' and they said, 'We are*
> *the Committee of Adjustment, sir. These men are missing, presumed dead*
> *… we've got to collect their equipment.' They had this blanket spread on the*
> *floor and our stuff all mixed together. So I said, 'That's mine, that's mine,*
> *that's mine. You can have the rest.' And I went back to sleep.*[7]

While working at E. K. Cole, Sydney Grimes had met Iris, a girl who was a temporary copy typist there, filling the time until her seventeenth birthday, when she could begin training as a nurse, like her father and mother. Sydney was only eighteen months older than her but, Iris says with a smile, 'he looked very grown up, so I always called him "Mr Grimes".'[8]

Grimes says he knew at once that she was 'the one', but at first he didn't want her to get too involved, 'in case I didn't make it through, but she wanted to get married in case anything *did* happen to me! I realised I didn't want to lose her and needed to commit.'

Iris was now working as a nurse at a hospital in Leytonstone and Grimes visited her whenever he was on leave.

We lived life to the full and in the moment – we never said it, but I suppose at the back of our minds was the thought we might not get those moments again. We didn't talk about what I did, I didn't want to bother her with all that. She was seeing the injured brought in all the time, so I felt she was taking more chances than me, to be honest, living near the docks which were being bombed all the time. She had enough worries of her own. I didn't have a lucky charm, but I flew with a picture of Iris tucked safely in my wallet. It meant she was with me every time I got airborne. I never did tell her about that; in fact I'm not even sure she knows today!

It was true – amazingly – seventy years after their war together, his beloved wife *still* did not know her picture had kept him company on every operation over enemy territory.

It is not difficult to understand their reluctance to talk about the dangers they faced. Iris recalls:

I tried to be brave when I was with him, and not show him I was worried, but it was really horrible when he had to go back to the squadron. I would lie awake at night, wondering where he was and what he was doing. I knew he was in danger and I thought about him being killed, though I didn't tell him of course. I'd hear the bombers going over and wonder if he was in them, wonder if he was safe, wonder if he'd get home.

Iris herself was far from safe:

We were being bombed in east London, which was very frightening at times – the Germans were doing it to us and our boys were taking it back to them. I had a friend who was killed in the bombing but I was seeing the reality of the war constantly with the casualties being brought into the wards. It made me think about what the Germans were suffering too. I'd had a German pen-pal before the war and we would write and tell each other about our lives and family. I used to love receiving her letters and news; we knew everything about each other's lives. I worried about her, wondered what she was going through. Luckily she survived and we carried on again after the war was over!

Sydney and Iris Grimes's wedding

As Sydney and Iris relived their wartime experiences from their retirement home near Cambridge, Sydney's eyes lit up. 'I've just remembered!' he says:

> *I did once talk about death with my crew. I've only just remembered this incident! Before we had even done one op on 106 Squadron, our navigator, Bruce Bayne, had to stand in for another nav who was sick. So his first op was to be with another crew. When we woke up that morning, he was lying in the next bed to me writing his last will and testament on the wall – I can still see him doing it now! I asked what on earth he was doing it for. He simply replied, 'Well, you never know, do you?' If I don't get back, you've seen my will, you can tell my family what my last wishes are.'*

Airmen had to be lucky as well as good, and Iris's husband readily acknowledges his own good fortune, especially after a low-level attack on Stettin in south-west Germany. 'All the way home, we managed to find every flak ship out there. We were weaving all over the place, trying to find a clear track, and ended up over the Isle of Sylt, a German naval base, so we got even more shot up there!' The ground crew counted over 100 holes in the aircraft when they got back and their Lancaster had to be written off. As flak battered the aircraft and punched holes in the fuselage, while the pilot threw the Lancaster around, it had sounded 'like handfuls of gravel being thrown against a tin shack' and Sydney could smell 'burning chunks of flak shell'. 'The wireless op wasn't a busy man,' he says. 'Most of the time you were just listening to the radio, so you had a lot of time to contemplate what was happening around you. But you couldn't show any fear, so you became a fatalist. You had to, or there was a danger you would end up LMF.'

★   ★   ★

LMF – lack of moral fibre – was the RAF's catch-all euphemism for a number of different circumstances, covering everything from becoming 'flak happy' (as combat fatigue was called) to refusing to fly on operations or failing to return to base after leave, and it could lead to disciplinary action. However, the basis for such charges was often extremely flimsy and there was a widespread – and possibly justified – feeling that charges of LMF were sometimes laid as much *pour encourager les autres* as to punish genuine cowardice. It was often almost impossible to tell whether a pilot had turned back because of mechanical failure or navigational error, or because he'd simply bottled it, but, in contrast to traditional British notions of justice and fair play, the authorities sometimes seemed to operate on a basis of 'guilty until proved innocent'.

An op to Leipzig flown by Bob Knights' crew almost ended up with LMF charges against them. They were flying through thick cloud on their way to the target when the airspeed indicator froze up, leaving Knights flying by 'feel'. Suddenly he noticed that they were losing altitude at a rate of 2,000 feet per minute. The flight engineer told him that the number-two engine had stopped. 'It can't have done,' Knights said, 'because I'd have felt the swing,' but that was explained when the flight engineer looked out of the other side, and saw that the number-three engine had stopped as well. 'So there we were,' Knights said, 'on two engines, going down, in cloud, so I yelled at the bomb-aimer to jettison the bomb-load.' They came out of the base of the cloud at 5,000 feet and eventually managed to restart the engines, but since they now had no bombs to drop, they turned and headed for home.

'We lost a lot of aeroplanes on that Leipzig raid and there was hardly any opposition,' Knights said. 'I think they iced up but nothing official was said.' However, when they reported what had happened at the debriefing, the Station Commander – a Group Captain, whereas Knights was a Flight Sergeant at the time – treated them with a scepticism that he might well not have shown to a fellow officer. 'They were looking for people suffering from LMF,' Knights said, 'and I was getting quite belligerent at the third degree I was getting.' Eventually, the Group Captain backed off and told them go to bed, and they heard no more about it.

In the vast majority of cases, an inquiry today would be much more likely to conclude that the pilot was suffering from PTSD rather than LMF. Australian pilot Bruce Buckham bravely spoke up at an inquiry on behalf of another pilot who was being accused of lack of moral fibre. Buckham freely admitted his own fears, saying that he was 'dreadfully afraid' every time he flew an op, but 'one of the things you had to do was not to show anybody else how you felt. Being shot up by ground defences or fighters, or coned by searchlights, this plays on a man's mind all the time, and after a certain number of trips it would only be human for a bod to think "There's no future in this," and you'd get the extreme cases where they'd either go missing or cave in.'

At the other pilot's LMF inquiry, Buckham told the panel of senior officers, 'This man has ... forced himself to go right into the heart of Germany, he's dropped his bombs and brought his crew back twenty-one times. Now he's reached his limit, he can't go again and if I had my way, I'd give him the VC because he's gone twenty-one times before this awful thing has happened to him. I don't think you should court-martial him.' His advice was heeded and Buckham later said, 'I've got as much respect for that bloke today as I have for a VC.'[9]

However, not every crew was as motivated and conscientious as those on 617 Squadron. There were tales of aircrew sent to a heavily defended target skirting the edge of the defences and dropping their bombs wide. 'An awful lot of bombs fell outside the target area because people just didn't press on,' Johnny Johnson says. 'And there were also instances where people aborted because they just didn't like the look of the target. Icing could sometimes be a problem, but there were times when it was just used as an excuse to come back. It did happen, though it certainly never happened with us.'[10]

The consequences of being deemed lacking moral fibre were draconian. One airman recalls the 'sad, sad sight' of a comrade being humiliated for having had enough. In his diary in June 1944, he recorded:

> *The entire strength of the station was on parade. By order. No exceptions. This sergeant had refused to fly an op. He had been accused and found guilty of LMF. There he was standing out in front, all on his own, in full view of every person in the unit, to be stripped of his wings and then his sergeant's tapes. They had all been unstitched beforehand so they came away easily when they were ripped from his uniform. He was immediately posted elsewhere.*[11]

Les Munro was brutally realistic in his view of the risks the crews faced:

> *There was certainly no discussion about what we'd do when the war was over – we never spoke of the future. There wasn't much point if there was a possibility you might not survive. I was a fatalist – whatever was going to happen would simply occur come what may. If I was killed, so be it. I never thought, Come tomorrow I might not be alive – it didn't worry me. If you started to worry if you'd cop it on any raid – that was the beginning of the end of it all. You'd end up being taken off ops, and no one wanted that.*

★   ★   ★

Sydney Grimes had completed his first tour in September 1943 and became an instructor at Balderton, where one of his first pupils was Leonard Cheshire, who arrived in November 1943 to do a crash course on Lancasters before taking over command of 617 Squadron. 'He was the finest man I ever met in my life,' Grimes says.

> *He was a group captain and he would sit with each of us on the aircraft, asking us about our role and some of our experiences. He sat down on the step beneath my feet and chatted away, genuinely interested in my job and what I had to do as part of the crew. He wanted to know about me, and my views on the job, the Lancaster, 5 Group – everything. I'd never had a pilot, let alone a senior officer, do that before. As he finished, he asked me if I wanted to go back on ops. I said, 'Yes sir, if the war is still going on when I finish here.'*

Within a few months he had joined Cheshire's new squadron. Bored with giving the same lecture three or four times a day, and wanting to go back to the front line, he joined another instructor, Bernard 'Barney' Gumbley, a pilot who was looking for a crew to go back on ops with him and had been offered the Pathfinders or 617 Squadron.

Gumbley was a New Zealander from Napier in Hawkes Bay. His parents were originally struggling small shopkeepers, and so poor that Barney and his brother slept in a tent behind the shop, except in the coldest weather when they would take refuge in the attic. However, in the early 1930s they set up a mobile motion-picture business, touring small towns and villages in rural areas of the South Island with a film projector mounted on the back of a Ford Model-T truck. They persuaded the townsfolk and villagers to cut a hole in the rear wall of their community halls and then reversed the truck up to the wall, pushed the lens of the projector through the hole and projected their movies onto a portable screen inside. Gumbley's brother painted posters of the movies on show, Gumbley himself manned the projector and his sister sold the tickets. The nitrate film stock used then was highly inflammable, and on one occasion the projector caught fire during a screening, forcing Gumbley to jump into the driving seat and drive the burning truck away from the building to stop the hall being burnt down. He was called up by the RNZAF in February 1941, and, after arriving in England, served a tour with 49 Squadron. Having completed his final trip of that tour, the fun-loving Gumbley and his crew 'celebrated at the local pub, winding up at a dance on the camp, and they say that I really came out of my shell and nearly disgraced the fair name of New Zealand!'[12]

Barney Gumbley's crew

Awarded the Distinguished Flying Medal, he took his Aunt Sophie and Uncle Ernest to the investiture at Buckingham Palace, and described what happened in a letter to his loved ones. 'We joined the queue fairly early at the gates,' he said:

and I don't know who was the most itchy. It was agony. Anyway, after a long wait we move through the courtyard where I left the others to wait in the hall while we paraded inside to be checked present and double-checked in case we decided to creep out. There must have been 1,000 of us altogether. First there were the Navy types, mostly with ribbons on their chests overflowing onto the other side of their tunics. Then the Army. Not so many of these. We came up next, about 100 of us, streams of civilians and Home Guard ... An hour and a half later, the queue began to move past 'his nibs'. Ages went by, then it was the Air Force's turn.

When his name was read out, Gumbley was:

in such a flat spin as I'm blowed if I heard it, but I stepped up and bowed (can you imagine it?) ... Having chatted for a few seconds, I stepped back, bowed again, marched off with medal swaying and before I had a chance to look at it, another type grabbed it and put it in a box. Pretty smart work I thought. The ceremony did impress me though, and the next time I know what to expect! King George stood the strain pretty well seeing that he does the same thing to literally thousands week in week out I suppose. And I have a whole lot of respect for him considering the impediment in his speech, which does make him a trifle self-conscious.

Grimes was still only twenty-two when they crewed up together, whereas Gumbley, a quietly spoken, shy man 'but very well liked', was much older than all his crew – twenty-nine and, says Grimes, 'an old man by the standards back then, but he was the complete pilot. He loved to fly above a perfectly flat cloud, put the wing down perpendicular, and carve a figure of eight in the cloud!'

Grimes had no hesitation in joining up with him, but he 'wasn't keen on going back to Main Force with all of the losses and mass raids', so 617 Squadron seemed like a good idea. 'I felt we were joining an elite squadron and it was an honour to go there.' The procedure was so informal that he arrived on the squadron without any paperwork being completed: 'I just turned up and started flying! I actually did two ops to the *Tirpitz* before I'd become an official member of the squadron.'

Grimes had an extra motivation as he prepared for the first of those ops, 617 Squadron's second raid on the *Tirpitz* on 29 October 1944. 'My brother was in the Navy on HMS *London*, escorting Arctic convoys,' he says. 'He'd told me about the *Tirpitz* and I knew it worried him, so I thought if I could do something to help him out, all the better.'

It proved to be a real baptism of fire. 'The flak was very heavy on that trip,' flight engineer Frank Tilley says:

You could see it coming up, coloured shells – yellow, red, black and grey – rising slowly towards you, so close you felt you could almost reach out and touch them. Then they exploded and blossomed out in a cloud of smoke and shrapnel. As you fly towards it, it's like flying into a tunnel of darkness. If it hits you, you get peppered with showers of shrapnel. Of course, if one hits you, there's nothing you can do. You realise it could be the end, but you just have to sit it out. You just have to hope that it hasn't got your name on it.[13]

Unfortunately some of the flak turned out to have Bill Carey's name on it. From Mount Gambier in South Australia, Carey – 'a very charming chap', according to Basil Fish – had been born in the dying days of the First World War. He enlisted with the RAAF in August 1941 and left for England twelve months later for advanced training.[14]

After making the conversion to Lancaster bombers, Carey was posted to 106 Squadron in February 1944. He flew ops over Stuttgart, Nuremberg and Berlin, as well as the regular Main Force attacks on 'Happy Valley' – the Ruhr. After completing his first tour, he joined 617 on 10 April 1944, making his 'second dickey' flight (the training flight as an observer required of all pilots before they could take the controls themselves) the same night on a raid on St-Cyr in occupied France. His first flight with his own crew was the attack on the marshalling yards at Juvisy a week later.[15] Short and stocky, with a habit of weighing the most casual of remarks as if it might contain a deadly insult, he was known on 617 as 'The Australian James Cagney', which may well have been a reference to his lack of inches rather than his acting ability.[16]

As Carey made his first, unsuccessful, run over the *Tirpitz* on that crisp October morning, flak bursts from the ship's guns riddled his aircraft, destroying his starboard outer engine and causing a fuel leak, but although fuel was 'spraying from it, the damaged engine did not catch fire'.[17] Alex McKie, navigator in Carey's 'Easy Elsie', had guided the Lancaster on to its bomb-run and was watching the *Tirpitz* 18,000 feet below when suddenly 'the aircraft heaved amidst a massive explosion' and he muttered to himself, 'Christ, that was close.'

They were thrown off course and he heard the Canadian bomb-aimer, Don McLennan, call, 'Dummy run, we have to go round again.'

'For God's sake Don,' McKie shouted, 'let's drop the bloody bomb and get the hell out of here!'

Carey's voice broke in, cool as a cucumber: 'Steady Mac, we haven't come all this way to waste a bomb. That son-of-a-bitch battleship is going to pay for hitting us on the first run!'

They actually had to make another six runs before, with flak bursting all around them, they got the bomb away. 'It was hell on earth for us,' McKie said, 'but we did it anyway.'

As Carey attempted to make a low-level getaway, he was hit by a further burst of flak, fired by a lone gunner in a village on the small island they were overflying; it silenced his port inner engine and holed his fuel tanks.[18] Gerry Witherick, the rear gunner, saw fuel 'streaming past my turret like nobody's business'.[19] By a miracle none of the leaking fuel ignited, but the flak had also damaged the hydraulics, so that the landing gear and flaps lowered and the bomb doors swung open, causing a heavy drag on the aircraft. With only two surviving engines – both running at maximum revs to counter the drag from the flaps, bomb doors and landing gear – and severely depleted fuel, Carey felt they had no chance of recrossing the North Sea, and, in rough seas, ditching in the hope of being picked up by a destroyer was not an attractive proposition. Instead, he opted to try to make for neutral Sweden and crash-land there. Sydney Grimes had heard Carey radio that he'd 'lost one engine and another was dodgy. He was telling Tait he wouldn't be able to get back. Tait just wished him luck and gave him permission to leave.' However, Frank Tilley 'didn't see Carey get hit, or even hear that it had happened. That's just the way it was – you didn't really hear what was happening to others.'

'Unfortunately, we had now lost an engine, hydraulics and quite a bit of fuel through the various holes in the aircraft,' McKie said. 'We were not going to get home, that was for sure, so our options were limited. Crash-land in Norway and be taken prisoner – not my idea of fun. Ditch in the freezing sea and almost certainly die – again, not a brilliant choice. Or limp inland as the fuel bled away in the hope of finding a field to put her down in. I grabbed my maps and started working.'

Carey told his crew, 'We can't ditch and we can't get home, so it's over the mountains to Sweden.'

Gerry Witherick was aghast. 'It will ruin my reputation. I always get home. This can't happen to me.'

'Can't it?' Carey said. 'You watch!'[20]

He set his aircraft to climb, but the wounded Lancaster was agonisingly slow to respond and he ordered the crew to reduce weight by dumping every movable item, including their radios, parachutes and bombsight, though in case of attack by German fighters he told Witherick in the rear turret to retain his guns until they had crossed the Swedish border. They destroyed the top-secret equipment, such as the Gee navigation set and the SABS, with a few blows from the fire-axe before dumping them out of the hatch, and threw their maps and other classified material out of the bomb-bay.

Carey nursed 'Easy Elsie' on for more than 200 miles, inching upwards to gain the height needed to clear the forbidding Norwegian mountain ranges barring the way. He breathed a sigh of relief as the altimeter reached 6,000 feet – the minimum height necessary to ensure a safe transit. They crossed into Sweden through a high mountain pass with 100 feet to spare, and, after his crew had used emergency air-bottles to lock down the undercarriage, they crash-landed in a marshy field near the Lappland village of Porjus, deep inside the Arctic Circle.

By a miracle, everyone survived the impact, the only serious injury being Carey's knee, 'cut to the bone' and badly dislocated as they crashed. The crew gingerly extracted their skipper from the wrecked Lancaster and Witherick bound Carey's knee while the others covered him with parachutes and lit a fire. They also set fire to their aircraft, destroying the rest of their maps and equipment.

'We found out he was missing after we got back to Lossie-mouth,' Sydney Grimes says.

I don't think we really thought all that much about it. We accepted we were in a dicey business and so people would get killed or go missing. You were just thankful it wasn't you – it was another bullet we'd dodged – wrong place, wrong time. After he was shot down, the Committee of Adjustment just came to clear his kit away and that was that, he'd gone. Of course, at that point we had no idea what had happened to him, if he was alive or dead, but someone soon came in to take his bed space. Such is life!

Carey and his crew were still very much alive, and had been picked up by the local Home Guard. After a brief internment by the Swedes, they were sent back to the UK. Preserved in the deep freeze of the Arctic tundra, the wreckage of 'Easy Elsie' still lies where it fell, mute testimony to the skill, dedication and courage of Carey and his crew.

★   ★   ★

The remaining Lancasters had continued the attack, and though the first two Tallboys missed the target, Jack Sayers, who dropped the third, saw the vivid flash of a bomb burst. Although the cloud cover made the sightings problematic, several other pilots also claimed hits or near-misses. Bob Knights, flying the last of his seventy combat ops, was circling around while his bomb-aimer and navigator plotted the fall of all the bombs. He said the *Tirpitz* 'rocked' under the impact of one blast, and he saw gouts of black smoke coming from the starboard bow and brown smoke from amidships.[21] Several crews made repeated passes over the target in the hope that a break in the cloud would allow their bomb-aimer a glimpse of the ship, and most eventually dropped their bombs, though three took their Tallboys home with them rather than drop them blind.

Mac Hamilton had a 'hang-up', meaning his bomb would not release. He went round again and again, with Willie Tait flying alongside to draw off some of the anti-aircraft fire. Hamilton finally managed to release his Tallboy on the fourth circuit, by which time they'd found enough of a break in the cloud to sight the target and convinced themselves that they had scored a hit on the *Tirpitz*. Hamilton was the last to leave the target but the returning crews still had other dangers to face.[22]

Bruce Buckham in the Lancaster camera plane had to crash-land at Waddington. Halfway across the North Sea he heard 'a terrific crash, and in the morning light we could see the starboard leg was hanging down, engine nacelles flapping in the breeze and a gaping hole in the wing. A shell had passed through the under-carriage and between Number One and Number Two fuel tanks.' They landed at Waddington on one wheel after a marathon flight lasting fourteen hours and twenty minutes.[23]

Joppy Joplin's crew also had a heart-in-mouth return flight, with an over-revving port engine eating into their fuel. 'With this engine problem, we had a long and dangerous journey back to safety,' Frank Tilley says, 'a continuous trek at six hundred feet over never-ending water, nothing to see, nothing to look at, just water. It really was causing me some serious concern, because if we went down in that, our chances were almost zero. I really was worried about the fuel state and wanted to get it on the ground.' They managed to reach the Shetlands and landed safely on a Coastal Command airfield, despite its having only 'a very short runway with a small mountain at the end of it'.

Arthur Kell, with his 'stowaway' Aspro Astbury aboard, had spent so long circling over the *Tirpitz*, waiting in vain for enough visibility to bomb the ship, that he was desperately short of fuel and had to make for Sumburgh in the Shetlands rather than Lossiemouth. The Station Commander there ordered them to go back out to sea and jettison their bomb-load before landing. Landing with a Tallboy aboard was now routine for 617's pilots and Kell had no intention of wasting his precious and highly expensive cargo, so he flew out to sea, opened and closed the bomb doors as if he had jettisoned the bomb, then returned and landed. The huge weight of the bomb put such a strain on the Lancaster that there was a danger of permanent damage to the main spar if it was left in place too long, so, having taxied to the grassed dispersal area, Kell reopened the bomb doors and after checking that all the bomb switches were set to 'SAFE', Aspro released the Tallboy onto the grass under the aircraft.

Unfortunately the Station Commander, who had never seen a Lancaster before, had just come over for a closer look, and was now standing, ashen-faced, staring at the bomb. Aspro strolled over to him and patted him on the shoulder. 'It's quite all right, sir, never you fear,' he said. 'I dropped it "safe" and it's not going to go off. We'd have heard the bang by now if it was.' He banged the nose of the Tallboy for emphasis as he spoke. The Station Commander gave a weak smile, tottered back to his car and drove off without another word.[24]

Benny Goodman's crew had returned safely to Lossiemouth, and Benny was looking forward to a well-earned rest before his much-anticipated rendezvous with the 'very pretty young WAAF' he had bumped into the night before and who had promised to meet him after the op. He and his bomb-aimer, Tony Hayward, had seized the chance for a few hours' sleep, but he had told Hayward, 'Whatever happens when we get back from this op, no matter how tired I am, when we've had a bit of shut-eye you have to wake me up so I can meet this girl.' In the event Hayward woke up and sneaked out of their room, leaving his exhausted skipper sleeping soundly. He went off to meet the WAAF himself, who apparently was 'most welcoming'. 'I was absolutely bloody furious with him!' Goodman says, laughing despite himself.

★   ★   ★

Surveillance flights after the attack once more showed the *Tirpitz* still upright in the water, but the great battleship had actually suffered more severe damage. A near-miss from a Tallboy, exploding within 20 yards of the ship, had bent the port propeller shaft and buckled the plating around it, causing fresh torrents of seawater to pour into the damaged sections of the ship. Whether the *Tirpitz* still posed a genuine threat to the Arctic convoys was dubious in the extreme, for it had sustained so much damage in this and the earlier raid that it was now incapable of venturing onto the open seas at all, and German commanders saw it solely as a floating gun battery. However, it remained afloat, and, not for the first time, Churchill's personal obsessions were driving his country's military strategy, as he made increasingly strident calls for it to be sunk. 'No other target is comparable to it,' he had previously claimed, and he now made his wishes even more explicit: 'Every effort should be made to attack this ship, even if losses are incurred.'

Another joint attack by 617 and 9 Squadrons was duly authorised, but little time remained to carry it out. At Tromsø's latitude, inside the Arctic Circle, there was already only a handful of hours of daylight, and after 26 November the sun would not rise above the horizon until the following spring. Bad weather over the target caused repeated postponements of the raid as the 617 crews found themselves shuttling between Woodhall Spa and Lossiemouth, but on 12 November 1944, Operation Catechism, the third and final assault on the *Tirpitz*, was at last launched. The crews had once more assembled at Woodhall Spa on 11 November for their final briefings. Among them was Aspro Astbury, again determined not to miss the action. This time he had gone AWOL from a holding unit in Bournemouth, from where he was about to be repatriated to Australia, but he was unable to resist the lure of a final flight with 617 and the chance to be in at what he hoped would be the death of the *Tirpitz*.

John Leavitt's novice crew were also among those assembling for the briefing, and their first ever op was to be the attack on the *Tirpitz*. The wireless operator, Colin Cole, recalled:

> *We were very unusual in that we hadn't done any ops before being posted to 617 Squadron. I think we were all a little bit startled by this. I knew about the Dams raid from the news, but not that I was joining the squadron who carried it out. I can't even remember how we found out. It was never discussed on the squadron – I suppose it was old news. You didn't dwell on past glories, or even on the past; it was what was ahead that mattered.*[25]

Four years earlier, Cole had been a sixteen-year-old boy in his home town of Guildford, watching 'the Battle of Britain going on over my head. All my friends used to watch them wheel and dive above our heads, cheering them on.' One of his comrades on 617, Murray Vagnolini, the wireless operator with Nelson Hill's crew, had actually been evacuated from the blitzed East End to a 'safer area' that turned out to be not far from

Biggin Hill. 'So we'd actually been evacuated to the centre of the Battle of Britain! Oh, it was fantastic watching the Battle of Britain there! We had a Spitfire crash-land in the field just by my uncle's house.'[26]

Inspired by his own grandstand view of the Battle of Britain, Colin Cole joined the RAF as soon as he was old enough, but did his training in the sleepy setting of Yatesbury in Wiltshire. 'We'd seen the Battle of Britain and the Blitz. I can't imagine anyone had even seen a bomb at Yatesbury!' he says now, with a chuckle. After joining 617, there were months of training, familiarisation and bombing practice on the Wainfleet ranges, and they were not officially informed when they became operational. 'Our names just appeared on the Battle Order which was posted in the Mess. We were pleased to see that at last we had progressed onto ops, although looking back, it was an unusual op to have as your first trip.'[27]

Most of the other crews already knew the details backwards and had the flight plan firmly imprinted on their minds, but there had been one worrying development since the last attack. When briefing the pilots, Willie Tait had told them, 'I'm sorry to tell you, chaps, but since the last trip, the Germans have moved two squadrons of fighters to be near the *Tirpitz* at Bardufoss.'

That news sent a chill through every man present. Bardufoss was only ten minutes' flying time from Tromsø, and the thought of the German fighters swarming around their now almost defenceless Lancasters, with only rear guns to protect them, 'made us swallow a bit', Frank Tilley said, with a shudder at the memory. They would have swallowed even harder, had they known the calibre of pilot flying those fighters: the top three pilots had over 500 kills between them.[28] The loss of the mid-upper turret and gunner gave the Lancasters a little more airspeed, but they were also aware that 'we'd lost a pair of eyes ... scanning all the time. So we looked on that with some misgivings.'[29] 'We were a little bit apprehensive,' another crewman recalled, 'but, like good little boys, we all went off.'[30]

Not all the aircrew knew about the fighters. At least one pilot didn't mention the fact to his crew, who remained in blissful

ignorance as they prepared for the op.[31] They took off from
Woodhall Spa and some crews flew to Lossiemouth to refuel,
while the remainder went to Milltown. That evening, with a few
hours in hand before take-off in the very early hours of the
following day, some of the aircrew, still wearing their flying gear,
went to a dance in Lossiemouth village.

For Frank Tilley it was one of his first times staying away from
home base:

> *so there was a sense of doing something adventurous. The girls were pleased
> to see us, though I think they were quite used to seeing rough old aircrew
> turning up in flying kit, trying to impress them. It was a bit macabre I sup-
> pose. Here we were dancing, enjoying female company, having a beer, and
> the next day we were off to drop bombs on German sailors. Of course we
> didn't talk about it. You didn't reveal your feelings to others; you just got
> on with it and lived in the moment.*[32]

It started to snow that evening, and by the time they were preparing to take off at three the following morning, the snow on the
runway was two feet deep. 'It was a filthy night,' Tony Iveson
recalled, 'dark and cold, and the aircraft had hoar frost on the
wings. There we were on a bleak and pitch-dark Scottish airfield,
with a twelve-hour flight in front of us, over mountains into a very
difficult country, facing unknown conditions and maybe a waiting
fighter squadron – it wasn't like a Sunday afternoon trip on the
Thames!' The two aircraft ahead of Bruce Buckham's camera plane
got bogged down in the snow, and he 'had to taxi around them in
flurries of snow'. Only nineteen aircraft managed to take off.

At Milltown, conditions were equally bad. It was so cold that,
after taxiing to the end of the runway, the aircraft had to shut
down their engines to have their wings resprayed with de-icing
fluid because they were in danger of icing up. It was a pitch-dark,
moonless night, with cloud cover obscuring the stars. The crews
taxied in turn onto the runway, were given the green Very light
from the caravan and then roared off with the engines bellowing
on full emergency power as the pilots struggled to coax their

aircraft, with their monstrous load of fuel and ordnance, off the ground. The Milltown runway terminated right at the water's edge so 'it was a question of take off or swim'.[33]

Once airborne, they climbed to 1,000 feet and flew north in radio silence. Their planned route was to fly up the North Sea for several hundred miles before crossing the Norwegian coast, where there was a known gap in German radar cover – though it has been suggested that the Germans might already have repositioned their radar towers to fill the gap in their defences that 617 had previously exploited.[34] As they approached the Norwegian coast the sky was just starting to lighten at the approach of dawn.

Climbing to 7,000 feet to clear the Norwegian mountains, they then flew north up the border between Norway and Sweden, trying to avoid detection by German radar for as long as possible. Even with the burden of their mission on their minds, the beauty of the snow-covered mountains with their glaciers and peaks tinted by the Northern Lights made an indelible impression on the aircrews.

In the pre-dawn of the morning of 12 November, with just a sliver of daylight, 'a silver thread on the horizon', the Lancasters arrived at their rendezvous point over Akkajaure Lake in Swedish Lappland. They had all flown individually, without seeing another soul, but within a very short time they all appeared out of the black western sky – another remarkable feat of flying and navigation. 'It gave me quite a shiver down the spine to see that,' Tony Iveson said. 'These guys had flown for about five hours, through total darkness with rather primitive navigation aids, but they made it. It made me feel proud to be part of it all.'

After circling for a few minutes, Willie Tait ordered a Very flare to be fired, signalling them to set course for Tromsø Fjord. They began their run out of Sweden at twenty to eight that morning. This time, dawn broke with clear skies, no cloud cover and perfect visibility. 'I had a good feeling about this one,' Iveson said. 'We had the weather, the dawn was breaking, conditions were ideal and there was a sense of excitement at what we were going to do. It was a big task, a big target; it wasn't just dropping bombs on German cities.'

As they climbed to their individual bombing heights, the bomb-aimers had an unimpeded view of the target, visible from 30 miles away. 'It was a magnificent sight to see the large snow-covered island of Tromsø ... and *Tirpitz* anchored facing towards us.'[35] 'In the distance, the sun was just shining brilliantly over the ship, shimmering in the water.'[36]

From 15,000 feet, even though she was well over 800 feet long, 'the *Tirpitz* must have looked like a dinky toy to our bomb-aimer,' Tony Iveson said. 'It was a gin-clear sky, perfect for our purposes, but then a little thought came into our minds: it'll suit the fighters pretty well too ... I thought, I hope to Christ the fighters don't get near us. If I get an Me 109 up my backside with his cannon, there'll be very little left of me.'

As the Lancasters climbed, they revealed themselves to German radar, which picked up the first trace of the raiders at eight o'clock, UK time, when they were still forty minutes from the target. An alert was radioed to the battleship, which at once went to action stations. The Commandant, Captain Robert Weber, made an announcement over the ship's tannoy: 'Twenty four-engine aircraft, south-west, 100 kilometres away,' which 'filled the air with electrifying tension', though the ship's gunners felt full of confidence, having driven off the raiders with only minimal damage a fortnight before.[37]

At 8.09 the *Tirpitz*'s own radar detected the bombers, and an officer, Alfred Zuba, began tannoy announcements, counting down the distance, kilometre by kilometre, as the bombers approached. At 8.27, lookouts on the bridge sighted the approaching Lancasters for the first time. Gunner Klaus Rohwedder watched the raiders as they flew towards the ship and thought to himself: 'If it's the same aircraft and bombs that were dropped last month, we won't survive.'[38]

Soon afterwards the battleship's main 15-inch guns began firing at maximum elevation using shells with short-delay fuses, even though the raiders were still over 20 kilometres away. The ship shuddered with each firing of the guns. Each shell weighed over a ton and, said Frank Tilley, 'You could see them arcing up towards us. They seemed to rise up so slowly, but then there were great unfolding golden clouds' as they exploded around the loose formation.[39] 'One shell could probably have taken two Lancasters out of the sky together,'[40] but fortunately, although the German heavy guns and the subsequent anti-aircraft fire 'hit our altitude exactly "cock-on",' Tilley said, 'they were a little bit early; the flak was bursting a few yards in front of us.'

Crucially for the success of the op, the smoke generators on the ship itself were now defunct, and none had yet been installed on the surrounding shores and headlands to create a smokescreen to hide the ship. As he made his approach, the only smoke that the squadron's commander, Willie Tait, could see was that drifting upwards from the *Tirpitz*'s funnel on the easterly breeze. As they closed on the target there was sporadic anti-aircraft fire from guns lining the fjord while 'the *Tirpitz* ack-ack and gunfire from two flak ships on the fjord hotted things up.'[41]

Most of the Lancasters veered away to make their bombing runs from the planned direction, but Lofty Hebbard, the bomb-aimer in Joplin's Lancaster and, like his skipper, a New Zealander, decided to head straight in on the track they had been following and 'not bother with the briefed routing. I never did ask him if he had permission to do that!' Frank Tilley recalls with a smile:

> We ended up on the longest bombing run in history. The flak was very heavy again. It got so close that you thought you could reach out and touch it. We were in the midst of a flak storm, surrounded by puffs of smoke. It's a fine dividing line between death and survival. You just have to hope and pray you'll get through it. We were also expecting fighters, so my eyes were out on stalks looking out. And you also have to concentrate on what you are doing: the engines, speed and everything else. There was a lot going on![42]

Colin Cole could see the flak heading his way and exploding around his aircraft. The crew had a narrow escape. 'I collected some big chunks of shrapnel that blasted their way into the aircraft,' he recalls. 'There was a red-hot, fist-sized chunk lying on the steps near my position; I should have saved it as a souvenir!'[43]

The additional tanks were now empty of liquid fuel, but that merely increased the danger, for they remained full of highly explosive fuel vapour and it would take just one piece of shrapnel to turn the interior of the aircraft into an inferno. However, to the great relief of the Lancaster aircrews, there was still no sign of any German Focke-Wulf 190 fighters from their base at Bardufoss. They were only ten minutes' flying time from Tromsø and would have offered a far more formidable threat than the flak, but, although the 617 crews did not yet know it, they were to have the freedom of the skies over the battleship.

While the rest of the Lancasters were at their optimum bombing height of 13,000 to 15,000 feet, Bruce Buckham's camera plane had gone in at 6,000 feet, but 'this was too unhealthy so we descended to about 2,000 feet and isolated the guns lining the fjord. One of the flak ships became somewhat pestiferous so we shot her up a bit and she disappeared up to the end of the fjord; the other kept a respectable distance.'[44]

As the Lancasters began their bombing runs, each aircraft's bomb doors swung open, exposing the gleaming metal casing of the Tallboy suspended there. The jolt to the aircraft as the 5-ton weight was released told the crew that their Tallboy was on its way to the target, even before the bomb-aimer's call of 'Bomb gone'. Bruce Buckham, watching from his camera plane, said the bombs 'appeared to travel in ever so graceful a curve, like a high diver, heading with deadly accuracy towards one point, right amidships of *Tirpitz*'.[45]

Tait's bomb, the first one, was dropped at 8.41 that morning. Its bright green paint made it stand out vividly against the background of blue sea and black rock and 'you could see it go down like a dart.'[46] By the time it reached the ship, the bomb was travelling at the speed of sound and it had impacted the armoured deck before the crewmen heard the whistle as it fell. 'The bomb was so quick, we could not follow it with our eyes,' Klaus Rohwedder remembers. It was followed in rapid succession by several more.

The first wave of nine aircraft had all dropped their bombs within one and a half minutes – 'Just think of nine five-ton bombs coming down at you in the space of ninety seconds,' Tony Iveson said. Bob Knights saw the first four strike on or near the starboard quarter, starboard bow, port bow and port amidships, near the funnel, and claimed his own bomb missed the port quarter by only 10 yards.[47] Klaus Rohwedder confirms the accuracy of Knights' observations: 'The first hit was through the foredeck and it created a huge hole in the side. Then a bomb hit the funnel and went through the deck and exploded. The men moaned and howled.' The screams of the injured and dying, echoing around the dying ship, are something Rohwedder has never forgotten.

The bomb that struck amidships at its terminal velocity of over 690 miles an hour had drilled through the 5-inch toughened steel armour plating of the battleship's deck and exploded near the boiler room. An inferno of flame erupted from the deck and sent a column of brownish-black smoke billowing into the sky. The concussive blast wave from the bomb left most of the crewmen stunned. Klaus Rohwedder didn't even see the explosion, 'but I felt the quake. The ship made a half-metre jump. The next minute it began to capsize.' The explosion below decks had blown a 45-foot hole in the *Tirpitz*'s side, and as thousands of gallons of seawater poured through it, the ship began listing to port.

The awesome power of the Tallboys was vividly demonstrated by one bomb that missed the *Tirpitz* and the fjord and struck the shore of Haakøy Island. Even though it struck solid rock, the bomb gouged out a crater 30 metres wide and 10 metres deep. And, when falling into the waters of the fjord, even near-misses by bombs with the Tallboy's explosive power were devastating. The shockwaves buckled the ship's plating and generated walls of water like tsunamis that smashed into it. As one man who later examined the hull of the *Tirpitz* remarked, 'she had eleven inches of armour plate above the waterline and thirteen or fourteen inches below, but on one side the bow had been punched in like a tin can by a near-miss. If I had not seen it myself, I would have thought it just another tall story.'[48] The near-misses further increased the flow of water into the ship, and another direct hit and another near-miss lengthened the gash in the side to about 200 feet.

Colin Cole watched his crew's Tallboy in flight through the open bomb doors: 'It was an amazing sight to see it dropping away, tracking to the target.' Sydney Grimes had wanted to watch his crew's bomb all the way to the target too, and 'it was quite frustrating when the bomb doors were closed and we turned away. I really wanted to follow it down and see what happened!'[49]

By the time Arthur Joplin's Lancaster dropped its Tallboy, the smoke from the previous hits was beginning to obscure the *Tirpitz*, but he held the Lancaster in level flight long enough for the camera to record the impact of the bomb, then 'hauled us out of the firestorm. We were all more than happy to get out of the centre of that barrage!' says Frank Tilley. 'As we dived away down to sea level, I could see the *Tirpitz* had taken a terrible pounding and most of the big guns seemed to be out of action. I had the feeling it was curtains for her this time.'

His feeling was right. Deep below decks on the battleship, Alfred Zuba and his crewmates, slithering and stumbling on the steeply sloping deck, watched as their First Lieutenant snatched up the emergency telephone, called the bridge and asked permission to abandon ship. 'Since he has to shout to make himself understood,' Zuba said, 'we can hear everything clearly and thirty men are waiting for the answer. It can perhaps give light, daylight and life; or you can keep them prisoners in the dark on the lowest deck of the ship,' into which water was already pouring. The Lieutenant repeated aloud the order he had been given: 'Don't get out.' They knew what that might mean, but all still remained at their posts: 'We wait for our fate, enclosed in an iron space.'

By now the *Tirpitz* was completely shrouded in smoke, not from its usual smokescreen, but from the bombs and the fires raging on board. Following on from 617, the pilots of 9 Squadron had no clear view of the target and dropped their bombs into the heart of the pall of smoke. Deep inside the ship, Alfred Zuba and his crewmates could hear the sounds of fresh explosions above their heads, shaking the ship violently and increasing the angle of list still more. Almost every hit and near-miss was to the port side of the ship's centre line, further destabilising it and accelerating the list to port. Within four minutes of the first bomb striking the *Tirpitz*, it was listing at 40 degrees, and Captain Weber finally ordered the lower decks evacuated. Zuba rushed for the emergency exit, where a score of men were already queuing, each one fearing for his life but knowing that only one man at a time could make his way through the narrow shaft. 'So we stand there and wait, with the floor burning under our feet.'[50]

Unknown to them, as the ship's angle of list continued to increase, Captain Weber was already giving his last order: 'Save yourself if you can.' Only minutes before he had been urging his men to remain at their stations, showing how quickly the prospects of the ship remaining afloat had disappeared.

Eventually Zuba clambered up the shaft, ran through a pitch-dark hold and reached the gun deck, 'where something fearful awaits us'. None of them could climb up the smooth, slippery surface of the steeply sloping deck. They tried again and again but each time they fell or slid back. The shaft leading to the upper decks was out of reach 20 feet above them.

Water now surged in, 'gurgling black and oily', rising rapidly to their chests. As he heard fresh bomb strikes on the ship, Zuba, flailing with his arms, caught a hand-hold and dragged himself up. A comrade above him reached down a hand and pulled him up. He managed to cling to a pipe a few feet above the swirling water and after a couple of agonising minutes dangling by his hands, feeling the strength ebbing from him, he managed to get his feet onto a ventilator and stood upright on it. A voice shouted from the emergency exit shaft, but it was still over 6 feet away and the only way to reach the steel ladder leading through it was to jump and try to cling to it. If he failed, he would drop into the water and drown.

Alfred Zuba gathered his strength and launched himself. His flailing fingers closed around the bottom-most steel rung and he clung on. He hauled himself up through the hatch, then turned to help a comrade behind him. There was no one else and the water was rising so rapidly that it had almost reached them again. They clamped the hatch shut, then climbed on through the Mess, searching for a way out of their steel prison.

A few minutes earlier, Johannes Ullrich, one of the seamen who managed to escape before it sank, had paused for a moment, holding on to the ship's rail and looking at the icy water 70 or 80 feet below him. He glanced at his watch – it read 8.47 – and then jumped. He was in full uniform and had no lifebelt. Somehow the ship's guns continued to fire for another two or three minutes, but by 8.50 the list to port had reached 70 degrees. The guns were then silenced as a huge explosion blew the C gun turret, with its 14-inch steel plating and twin 15-inch guns, completely off its armoured mounting and high into the air. As Ullrich swam away from the ship he heard the explosion and watched as 'the roof of the gun turret disintegrated like paper in the air.'[51] The turret crashed down into the water alongside the ship, crushing some crewmen who were trying to swim to safety.

Bruce Buckham's camera plane had descended to 200 feet, its cine-cameras whirring as they recorded the action. As explosions ripped through the ship, 'the *Tirpitz* appeared to heave herself up out of the water. It was awe-inspiring – a huge mushroom cloud of smoke rising thousands of feet.' He had continued to fly close around the *Tirpitz* as the other Lancasters turned for home, but now decided to follow them, leaving the burning battleship and its towering pillar of smoke behind.

As he headed for the mouth of the fjord, there was another, smaller explosion from the ship, and his rear gunner called out, 'I think she's turning over.' Buckham turned back and this time flew in at 50 feet. He watched with bated breath as, at 8.52 that morning, *Tirpitz* heeled over to port, 'ever so slowly and gracefully', and turned turtle, exposing the red-painted hull and embedding its superstructure in the seabed. 'We could see German sailors swimming, diving, jumping, and by the time she was over to eighty-five degrees and subsiding slowly into the water of Tromsø Fjord, there must have been the best part of sixty men on her side as we skimmed over for the last pass.' That was their final glimpse of the ship before they flew out of the fjord and headed for home across the North Sea.

The massive flood wave as the *Tirpitz* turned turtle had helped to wash Johannes Ullrich and other lucky survivors to the shore, but many of the men clinging to the hull and floundering in the icy water of the fjord drowned or died of exposure. The ship was still ablaze, and burning oil spreading across the surface of the water was an added hazard for the terrified seamen swimming for their lives. A woman living on the shores of the fjord, Thordis Ryeng, had hidden in a barn with her children during the raid, watching the walls shaking 'like a ship on the water in a stormy sea' as the huge bombs detonated. When the bombing stopped they went outside and heard 'many screams and shouts' from German crewmen 'in the most terrible pain. Two words were screamed over and over again "Mutter" [mother] and "Hilfe" [help]. The ship was covered by smoke and the water seemed to be burning, and in this inferno, men were swimming.'[52]

★   ★   ★

A slick of oil was spreading across the surface of the water. Survivors, barely recognisable as human beneath the thick black oil, struggled through it, but many others were floating face down, drowning or already dead. Klaus Rohwedder had climbed onto the hull of the ship as it rolled over and then jumped into the oil-covered water. He does not know how long he was swimming through it. 'I just thought, now I shan't see father and mother again. But you're so scared, you can't really think.' He managed to cling to a piece of floating debris until he reached the shore of Haakøy Island, then sheltered in a barn with other shivering survivors until rescued by German sailors in a Norwegian fishing cutter.[53]

Inside the upturned hull, the emergency lighting came on automatically, but lasted only four hours, after which the survivors were plunged into darkness, straining their ears for any sound that might indicate help was at hand. Alfred Zuba and the others trapped inside the hull tried to remain calm: 'None of us wants to show the others what he fears.' As the hours passed, they found some stale bread, cigarettes, a bottle of cognac and a tin of coffee beans. One man began to panic and was hurriedly silenced. Another said it was his birthday, and many of them, says Zuba, 'must have thought, "Let us hope your birthday is not your death day as well!"' Another man declared that if he got out alive, he would get married at once. The survivors began talking about their plans for the future, the sound of their voices perhaps helping to mask the hiss of rising water pouring into their refuge. The radio operator kept asking, 'Is the water rising? Shall we get out of here?' He too was told to be silent. They consoled themselves that the water in the fjord was only 17 metres deep, so a part of the ship must still be above the surface.

Some of the crew of the Tirpitz

Suddenly they heard a knocking sound from somewhere: 'Bang! Bang! Bang!' They picked up a fire extinguisher and banged back. They shouted and were answered by another group of survivors trapped in one of the switch rooms, but after a while they became hoarse and the answering voices grew weaker and fell silent. After that, Zuba and his comrades just tapped out SOS: three short taps, three long, three short.

One of them said he could hear the hissing sound of cutting gear, but as the rest strained their ears, all they could hear was the rush of water. Another then used the fire extinguisher to smash through the metal wall of their compartment, battering it until it gave way, but beyond that was only the armoured steel of the ship's hull. Soon afterwards, they heard the hissing sound again, stopping and starting and accompanied by banging and crackling sounds. They resumed their banging with the fire extinguisher, spurred on as the water continued to rise, spilling through the hatch. Now they began shouting, 'Hurry up! The water is getting higher. We need help!'

The hissing sound was louder now and the steel wall faintly warm to the touch. A red dot appeared on the wall and smoke and sparks poured into the compartment. 'For us, these red sparks are the light of life.' Smoke and fumes filled the air. Eyes streaming, gasping for breath, they clamped wet clothing against their faces, watching the red line eating through the steel. It cut a small opening, but then it stopped. They heard voices receding and then deathly silence. The opening was too small for anyone to squeeze through. They shouted and screamed, pounding on the steel wall, but only echoes answered as the water continued to rise.

At last they heard the sound of men returning and the sparks began again. They watched as the red line advanced and eventually a piece of metal clattered to the floor and faces appeared at the hole. It was so cramped that some of them could only get through with the rescuers pulling them while their comrades pushed from behind. The passage beyond was only 40 centimetres wide, then they climbed a ladder and pushed their way through another hole, this time cut through the actual hull of the ship. As Zuba squeezed through it, he could see the sky above him: 'The stars are sparkling. I will never forget that moment.'[54]

Soon after the sinking, rescue parties, including some of the survivors from the wreck, had climbed onto the outside of the hull and begun painting marks where they heard the sound of people knocking from inside. However, their welding equipment couldn't cut through the 13-inch steel armour plating and they had had to wait for more powerful oxyacetylene torches to be brought. All the time the tide was coming in and the water was rising inside the ship, filling the compartments. They heard the faint sound of some of their comrades singing 'Deutschland über Alles' but then the voices fell silent one by one. They managed to speak to one man who told them there were another twenty in the compartment with him and begged the rescuers to help them escape the rising waters, but when the tide reached its highest point, his voice fell silent and he and his comrades all drowned. Zuba asked about the fate of a friend of his who, he was told, was still alive, but trapped with fifty others deeper inside the ship. The next day he was told that all of them had suffocated before they could be rescued.

Klaus Rohwedder was among 200 men who had reached the shore of Haakøy Island unaided or been rescued from the freezing water when the *Tirpitz* went down, and, like Zuba, another eighty-four men trapped in air pockets inside the ship were brought out alive over the following two days as rescuers cut more holes in the exposed hull. Attempts at further rescues were then abandoned, although local people claimed to have heard tapping sounds from the hull for another week before the battleship at last fell silent. At least 971 crewmen lost their lives.

The fury of German commanders at the failure to defend the jewel of their navy, a ship launched four years earlier in the presence of the Führer himself, saw the Luftwaffe commander at Bardufoss, Major Heinrich Ehrler, court-martialled and jailed for three years with hard labour, after an inquiry revealed a tangled tale of negligence and incompetence among the air defences. Ten officers from the anti-aircraft units were also imprisoned. Although radar stations had picked up the first traces of the raiders at 8.00 that morning, confusing reports suggesting that British aircraft had been spotted making for Russia well to the south of Tromsø, and of others well to the north, meant that the alarm was not raised at Bardufoss until 8.18, nine minutes after *Tirpitz's* own radar had detected the Lancasters and its commander had issued a request for fighter cover.

Flying in their familiar 'gaggle' formation, the Lancasters had passed due east of Bardufoss at 8.24, but it was not until 8.30 that the fighters, already scrambled in response to the alarm raised at 8.18, began to move off. They had to taxi the length of the airfield before they could take off, and were then held up even further as the runway controller made them wait to allow an incoming transport plane to land. 'We had to wait for a Ju 52 to land down the hill,' one of the fighter pilots, Kurt Schultze, complained, 'so there was another five minutes lost.'[55]

It was later claimed that the fighter force had not even been told where the *Tirpitz* was anchored, other than that it was somewhere in the Tromsø area, and, according to one of the *Tirpitz*'s officers, at least some of the fighters had flown to its old anchorage in Kaa Fjord at Alta. It has also been suggested that a German radar officer with secret pro-British sympathies may have delayed raising the alarm to Bardufoss, but in truth, the level of miscommunication, misinformation and general incompetence among the defenders makes it unnecessary to add treachery to the mix. Whatever the reason for the delays, when the last Tallboy was being dropped, the fighter squadron was still three minutes' flying time away, and by the time they at last arrived, the Lancasters were speeding for home at low level, the skies were empty and the pride of the German Navy was already at the bottom of the fjord.

A Norwegian boy was arrested for clapping his hands at the news of the *Tirpitz*'s sinking and, if more discreetly, most Norwegians were delighted at the news – 'It was the happiest day in Tromsø,' one said.[56]

After the war, the *Tirpitz* was broken up for scrap where she lay – a perilous operation because the ship was still full of oil and unexploded ordnance – but much of the more accessible cabling and pipework had already been harvested by enterprising Norwegians.

★   ★   ★

German fighter pilot Kurt Schultze and 617 Squadron pilot Tony Iveson met many years after the war, and once they had overcome what Schultze described with a smile as Tony's initial reluctance to meet 'a damn German pilot', they became good friends. 'We, as human beings, finally came through,' Schultze said, 'seventy years after the politicians did with us whatever they wanted and we just said, "Yes, sir."'

Tirpitz anti-aircraft gunner Klaus Rohwedder feels 'deep respect for the crews of those planes, because flying into a wall of fire and then getting your bomb on to the target – it takes something to do that.' But although he too gives great credit to 617 Squadron for dropping those massive bombs from 15,000 feet with such phenomenal accuracy, Kurt Schultze cannot forget that almost 1,000 Germans died in that attack. 'It gets to me now,' he said, wiping away a tear. 'I was pretty tough once, but now it's all gone.'[57]

By contrast with that horrific death toll on the *Tirpitz*, 617 suffered no losses, and once clear of the Norwegian coast and the threat of German fighters, the long flight home was 'pretty damn boring', as one of Jack Sayers' crewmen remarked, 'life became pure tedium.' There was little chat or exultation among the aircrews on the flight back to Britain. 'Coming back from a trip there was always a great feeling of it's over! But this time it was all pretty quiet on the way back,' Tony Iveson said. 'We were all tired, we had been airborne for twelve and a half hours, and after the excitement and tension leading up to the actual attack there was a natural quietness on the way back, a release from all the tension; we'd got away with it again! I think my bomb-aimer even went to sleep down at the front!'

However, Sayers' crew soon found a way to enliven the journey:

The purists would have been horrified if they could have seen inside our aircraft. Vic Johnson, the engineer, was at the controls and Ernie Weaver and Jack Sayers, the bomb-aimer and pilot, were in the bombing hatch with the cover removed, popping away with Smith and Wesson six-shooters at startled petrels and albatrosses. The birds were fairly safe from these inexpert marksmen, but the North Atlantic was subjected to a terrible pounding![58]

The 617 Squadron crews landed back at Lossiemouth or one of the neighbouring bases between 2.45 and 5.00 that afternoon, and by six o'clock they were in the Lossiemouth Mess. Although they were all dog-tired, they decided to have 'a bit of a party' but, knowing that they still had to fly back to Woodhall and 'couldn't get too sloshed', they contented themselves with just a few celebration beers.

At eight that evening they received the news from the recce Mosquito that the *Tirpitz* was upside down in Tromsø Fjord. 'There was a cheer,' Sydney Grimes recalls, 'but the biggest emotion was a great sense of relief we wouldn't have to go back again! Our luck at not being attacked by fighters surely couldn't last!' 'Of course the party really got going then,' another aircrewman remembered, 'but after a short time I felt so tired that I went off to bed and slept for ten or twelve hours solid'.[59]

They flew back to Woodhall Spa the following day. 'As soon as we stepped out of the Lanc next evening at Woodhall Spa, we were besieged by photographers,' Arthur Ward, Willie Tait's wireless operator, remembers. 'Cameras clicked left, right and centre (incidentally none of those photos came out, it was too dark). When we reached the control tower, the funniest happening of all, the Station Commander had roped in the Staffordshires' Regimental Band and they were all formed up, blowing their heads off when we drew up in Willie's car.'[60]

The serious partying then began, and it also became a send-off for Keith 'Aspro' Astbury, though he first had to sit still for a roasting from the Station Commander, Group Captain Monty Philpott, after he discovered that Aspro had again gone AWOL to join the attack on the *Tirpitz*. With that formality out of the way, the aircrews sent the ever-popular Aspro on his way on a tide of beer and bonhomie. For the next few months aircrews could always get a rise out of Philpott by claiming to have spotted Aspro in some far-flung corner of the base or the Petwood Hotel.

John Bell had left the squadron before the *Tirpitz* op, but he'd returned for a visit just after the final raid that sank the battleship:

*I really wanted to get back and see the chaps. I suppose I was missing it all
really. They were all in the Petwood and it was great to catch up with them.
They were on a real high and I was really sorry I'd missed it. I suppose I had
a loss of purpose; I missed the squadron, the camaraderie, the action, the
buzz, just being involved with the group. I was no longer part of the battle,
but life had to go on.*

Bruce Buckham's camera plane had had enough fuel left after
filming the sinking of the *Tirpitz* to make a direct flight back to
his base at Waddington, where he arrived after another fourteen-
hour flight. His debrief was conducted by 5 Group commander,
Air Vice Marshal Sir Ralph Cochrane, in person. Undeterred
by Cochrane's seniority, Buckham's opening statement was the
informal but accurate comment, 'Well, we won't have to go back
after this one; *Tirpitz* is finished.'[61]

As soon as he received the news, Winston Churchill wrote to
Stalin, 'RAF bombers have sunk the *Tirpitz*. Let us rejoice
together.' At first, the reaction of some of the 617 Squadron crews
to the news that the *Tirpitz* had finally been sunk was rather more
matter-of-fact. 'We were pleased, of course,' flight engineer Frank
Tilley said, 'it meant we wouldn't have to go back again,' but he
remembers no big celebration beyond the usual post-op beers.
Tony Iveson thought it 'a special target and we were congratulated by everyone, the King and Eisenhower and Harris. We felt
we had really achieved something, done something that many
others had tried to do over about three years – we finished *Tirpitz*
off. It was another reason to be proud.'

Like his predecessor, Guy Gibson, after the Dams raid, 617's
commander, Willie Tait, was summoned to London for a press
conference and an interview with the BBC, as the Dambusters'
latest triumph made fresh headlines worldwide. Bruce Buckham
was also interviewed. That night, as every night, his wife Gwen,
back in Sydney, was listening to the BBC World Service. The
headline report was of the sinking of the *Tirpitz* and fears that
some aircraft were missing. 'I didn't know if Bruce was on that
mission or not,' Gwen Buckham later said. 'We never knew.'

Then, at the end of the bulletin, she heard her husband's voice as he was interviewed about the sinking. 'That was the only time I knew that he was still alive,' she said. 'It was the first time I had heard his voice in almost two years.'[62]

The rest of Tait's crew decided to escort him to London, and after a celebration dinner they left Willie at the Savoy with his mother and fiancée and, 'gatecrashed a few nightclubs … After an argument at the door we were made guests of [bandleader] Harry Roy at the Milroy Club. We got awfully tight and sang our squadron song with the band.' They went to bed at 4.30 a.m. but had to be up again two hours later to drive back to Woodhall Spa – 'two hours' sleep after a night like we'd had!' as one of Tait's crew complained[63] – to be congratulated by the Air Minister, Sir Archibald Sinclair. Only when both 617 and 9 squadrons were paraded for the ceremonial visit did Frank Tilley realise that 'we'd done something newsworthy, perhaps something remarkable. Here was a VIP congratulating us on our job. It was a bit strange really; we hadn't really thought about "pride" before then.'

'There was plenty of congratulations and backslapping,' Sydney Grimes says, 'but the best thing was that after his speech, Sir Archibald Sinclair gave us forty-eight hours' leave, so I managed to get home for Iris's twenty-first birthday.' Frank Tilley and his navigator Basil Fish spent their leave in London. They went to see the famous wartime comedian Tommy Trinder at the Palladium and when Trinder cracked a joke about the sinking of the *Tirpitz*, 'we just sat still,' the modest Tilley says. 'We didn't want to draw attention to ourselves. But when we told the story back on the squadron, Willie Tait's rear gunner said, "You bloody fools, you should have told the management and they would have had you up on stage!" but we didn't want anything like that!'[64]

Back at Woodhall Spa, there were more celebrations on 23 November as Bill Carey, who had crash-landed in Swedish Lappland after the second *Tirpitz* raid, and had at first been feared dead, returned to the base.

Bruce Buckham was among many aircrew decorated after the raid. He was awarded the DSO and his five crew received the DFC. Bobby Knights won a DSO, Pilot Officer Norman Evans a DFC, and Willie Tait – despite his later, characteristically self-deprecating view that the sinking of the *Tirpitz* had 'not contributed much to the Allied victory'[65] – was recommended for a Victoria Cross. In the end he was not given a VC – the top brass may have felt that after VCs for his predecessors, Guy Gibson and Leonard Cheshire, 617 Squadron had already had its fair share – but he was awarded a third Bar to his DSO; he also had a DFC and Bar.

The *Tirpitz* raid had been Tait's ninety-eighth operational flight. He flew two more to complete his century and was then required to step down from the squadron he had led with such distinction. Larry Curtis had already been stood down by then. His thirty-eighth and final op with 617 Squadron had been the sinking of the *Tirpitz*, and there could not have been a better note on which to finish. He had flown over seventy bombing ops and – unusually for wireless operators, who often missed out on the medals routinely awarded to pilots, bomb-aimers, navigators and gunners – had been rewarded with a DFC and Bar.

In September 1945, Tait visited Tromsø Fjord. It was the site of his most celebrated achievement but his tone was anything but triumphant. About to be broken up for scrap, the rusting hulk of the *Tirpitz* was, he said, 'huge, hideous, and stank like a charnel house. There were nearly a thousand bodies still inside the flooded hull and the treacly black fuel oil still seeped out of the rents ... this rusty tomb was nothing to gloat over.'[66]

Willie Tait's replacement, John E. Fauquier, was a much-decorated Canadian who had learned his trade as a 'bush pilot' in the Northern Territories of Canada and was beginning his third tour of bomber operations when he joined 617. Like Leonard Cheshire before him, Fauquier had also taken a drop in rank in order to return to combat operations. His reputation as 'a tough, hard Canadian' had preceded him and, said one 617 crewman, 'we had awaited his arrival with some trepidation; but nothing changed on 617. If you dropped a clanger you got hit hard, but that's exactly the way it was under Tait, too.'[67]

When he was awarded the first of his DSOs, Fauquier's citation stated that 'he set an example of the highest order'. Barney Gumbley could verify that. 'Our new boss is a Canadian,' he noted in a letter home, 'and heck he has started with a bang. He uses the big stick to good effect. He is "training" minded and gets us airborne regularly as clockwork. A good thing. It only remains to be seen whether his dictator methods show a dividend.'[68]

However, the men of 617, always notoriously relaxed when off duty, were horrified to note that the example that the tough Canadian intended to set for his new squadron included a course of compulsory early morning PE. After a few days of half-hearted cooperation, it was a great relief to them all when they were told that they could forget about the PE and get back on ops.

CHAPTER 14

The Last Christmas?

As the winter of 1944 began to bite, Germany's position grew steadily weaker. As the Soviet advance continued on the Eastern Front, Hitler abandoned his wartime HQ, the 'Wolf's Lair' in East Prussia, on 20 November and retreated to his Berlin bunker, from which he was never again to emerge. The disappearance of even the remotest possibility of a German landing in Britain led the Home Guard to be stood down on 3 December, and though the Battle of the Bulge began in the Ardennes on 16 December and raged for five weeks, causing 100,000 casualties to both sides, it would prove to be Hitler's last real throw of the dice.

For 617 Squadron, the air battles were far from over, but there was a genuine hope among the aircrew that Christmas 1944 would be the last of the war, though any festive cheer was diminished by the events of 21 December, which are seared into Frank Tilley's mind. The operation he and his crew embarked on was a combined 617 Squadron and Main Force raid on a huge syn-thetic oil refinery and fuel storage depot at Politz, near Stettin, in north-western Poland. The plant converted bituminous coal into the aviation fuel used by the Luftwaffe. Sixteen 617 Squadron aircraft were to take part in the raid and Tilley was flying with his regular crew in Lancaster T-Tare, piloted by New Zealander Arthur 'Joppy' Joplin, who had just celebrated his twenty-first birthday.

Joplin had always wanted to fly, though the realities of warfare had passed him by. 'My father and uncles had all fought in the Great War,' he says:

Indeed, things may have turned out very differently for me, as my mother's then fiancé was killed flying in the RFC! She then met my father after the war. But no one really talked about their experiences and I certainly never considered the dangers. And when the second war started the whole of the British Empire was wound up in the war – it was the way of the world, everyone was in favour of the war effort. Joining up seemed natural.[1]

After training, Joplin's was one of only a handful of novice crews sent straight to 617 Squadron. 'I have no idea why they chose us, I suppose we must have impressed someone during training!'

Arthur Joplin's crew

On the morning of the Politz raid, Joplin's regular bomb-aimer, Loftus 'Lofty' Hebbard, was in sick quarters with a heavy cold. When Flight Lieutenant Arthur Walker, one of the most experienced bomb-aimers on the squadron, heard about it, he went straight to see Joplin. Arthur needed one final trip to complete a double tour of forty-five operations, so he lobbied Joplin to take the ailing Hebbard's place. 'At the time, it seemed totally reasonable to try to get an extra op completed,' remembers Joplin. 'And I was very pleased to have such an experienced replacement, so immediately agreed. Looking back now, it's a decision I've always worried about.' 'Arthur wanted to do that final trip to get his forty-five,' Basil Fish says. 'I suppose if it had been me, I'd probably have wanted to do the same thing.' Frank Tilley felt the same: 'At the time, I didn't really think anything of it,' he says. 'It didn't seem strange at all back then. I just remember he seemed very pleased to be doing his last trip with us.'

All the NCOs like Tilley were housed in thin-walled wooden sheds or Nissen huts, and he remembers that the weather when he woke up on that December day was 'absolutely filthy, raining, very cold and foggy. A terrible day.' Spirits rose a little as he and the other NCOs ate their pre-flight breakfast in the Sergeants' Mess. On a normal day it would have been porridge, but as they were flying, it was bacon, egg and chips with plenty of hot tea.[2]

After the final briefing, they were taken to their aircraft and fired everything, testing the engines ready to taxi out, but the fog then thickened and the operation was cancelled. They shut everything back down, clambered out and returned to the Flights. Some hours later, even though the fog appeared to be little better, they were told that the op had been reinstated. 'I wasn't very happy when they said it was all back on,' Tilley says. 'I wondered how the hell we were going to get back on the ground afterwards.'

Sydney Grimes was also 'a bit apprehensive, but I was every time an op was cancelled and then reinstated. There was a feeling that this was no longer "neat and tidy". It had been scrubbed for a reason and there was a sense of chancing your luck by starting it all again.'[3]

The crew bus took Joplin's crew back out to their aircraft, and they went through their pre-flight preparations again and took off as normal, though it was pitch dark and so foggy that, says Tilley, 'we just disappeared into the mist'. The op itself was routine. They were not intercepted by night-fighters and, despite poor visibility, some confusion over markers and 'the usual flak over the target', the refinery sustained some serious damage and only three Lancasters out of a total force of more than 200 were shot down. Mission completed, 617 and the other squadrons turned for home.

However, the weather in England had deteriorated even more while they had been carrying out the op, and fog was blanketing

the country. Benny Goodman was told to try to get into Woodhall
Spa, where they had just fitted the 'Drem' airfield landing system,
named after an RAF airfield in Scotland – shrouded lights on ten-
foot poles, hidden in hedges and bushes to make them hard for
enemy reconnaissance aircraft to spot. The lights were only visible
to aircraft in the circuit around the airfield, and could be dimmed
to hide them from enemy aircraft. A returning pilot had to fly
round the circuit until he saw the lead-in lights – the flare path
leading to the runway – which were only visible once an aircraft
was on its final approach. It was a brand-new system and Goodman
had never used it before. 'But it was that or nothing,' he says. 'All
I could see was the glow through the fog, not the runway itself,
but I managed to get it down. Once we landed, it was so foggy I
couldn't actually see in front of me to taxi anywhere, so I just
pulled off to one side of the runway and shut the aircraft down!'[4]

When Joppy Joplin's crew reached the coast 'it all started to get
a bit grim. The fuel state on our aircraft wasn't all that marvellous
and everything was clamped down with thick fog.' In such conditions, Drem or no Drem, he felt there was no possibility of finding,
let alone landing at Woodhall Spa, and he decided to make for
Ludford Magna instead which, though it had a much shorter
runway, at least had a FIDO – Fog Investigation and Dispersal
Operation – system: twin pipelines alongside the runway through
which petrol was pumped and ignited, creating walls of flame
whose heat dispersed the fog and whose light could be seen from a
considerable distance … in theory at least. 'You were confronted
with two rows of flames either side of the runway which effectively
lifted the fog,' Frank Tilley says. 'The pilot flew into this tunnel,
apparently of clear air, but it was quite spectacular. It was like flying
into hell!' However, on this night the fog remained almost impenetrable – so thick that they 'couldn't see past the nose or the wing
tips' – and the light of the flames was almost completely obscured.

However, there were 200 other bomber crews airborne that
night and looking for a place to land, and when Joplin radioed to
get permission to land, 'so were what sounded like about a hundred others. The air was just jammed with people desperate to

land,' Tilley says. 'It was very worrying to hear all this going on. We knew things were looking bad for a lot of people.'

They no longer had enough fuel to divert to a Scottish airfield, where visibility was much better. Getting lower and lower on fuel and with no communication from the ground, they kept circling around Ludford Magna, when suddenly there was 'a terrific bang' and when Tilley shot a glance out of the port side, he saw that the wing just outside the outer engine was bent upwards by 90 degrees and knew at once that they were in serious trouble. 'The engine gauges were spinning back and forward and the aircraft was juddering and shaking,' he says, 'and the revs were up and down all over the place.' Joplin yelled: 'Full power! Full power!' and Tilley 'pushed everything through the gate' but nothing happened. 'I remember radioing "T-Tare crashing! T-Tare crashing!"' says Joplin. 'And I just managed to tell the crew to get to their crash stations. After that, I remember nothing. It all went black.'

Whether the bomb-aimer, Arthur Walker, heard the order is not clear, but instead of making for his designated ditching station at the rear of the aircraft, he went forward into the bomb-aimer's compartment. 'I knew we were going down,' Tilley says. 'Basil [Fish, the navigator] and I headed for our ditching station between the main spar and the rear spar. I dived over the spar and the next minute we hit the ground with a terrific bang. I was very lucky to get over just as we impacted. If not, I would have been catapulted clean out the front of the aircraft and certainly would have died.'

They heard the prolonged squeal of tortured metal as the aircraft ground to a juddering halt. The thunder of the engines ceased abruptly, but a moment later there was a whoosh! and a roar as the spilling fuel ignited. Tilley remembers seeing the flames as he tried to get up, but he couldn't move at first because his parachute harness had snagged on something. Fortunately he didn't panic and twisted and hammered at the release buckle until it came free. As he got up, he 'wobbled and collapsed again' and realised his leg was broken. Holding on to the fuselage, he dragged himself back to his feet and saw his friend Basil standing dazed and clutching his head. Tilley brought him to his senses, shouting, 'Come on Basil, we've got to get out of here!'

As the aircraft filled with smoke, Tilley groped his way to his designated escape hatch in the roof and freed it. 'I have no idea how I climbed out the hatch,' he says. 'My broken leg didn't hurt at that point, it just felt spongy and wouldn't respond, but I knew that I'd die if I stayed in that burning aircraft, so survival instincts took over. I had to get out. I had to get up a couple of times, hop towards the hatch, and somehow pull myself out.'

He slithered and fell over the side of the fuselage, crashed to the ground and began to crawl away from the burning aircraft. 'To my shame,' he says, 'I didn't spend much time checking on the others. I know I couldn't have done anything with my broken leg, but it still plays on my mind to this day. I still wonder if I could have done more.' As he recalls the incident seventy years later, sitting in his home, surrounded by RAF memorabilia, his eyes fill with tears and he falls silent.

Basil Fish had been knocked unconscious by a blow to his head as the aircraft crash-landed. 'I remember coming to with my feet trapped and blood gushing from my head,' he recalls. 'I was trapped in a burning aircraft and any man who says he wouldn't be scared in those circumstances is a fool or a liar! But what was I to do? You have to do something, you can't just sit and wait.'

Groggy and with blood pouring from his head wound, he struggled to extricate himself from the aircraft, and was violently sick as soon as he got out. 'The scene around me was just unreal,' he says, 'a burning aircraft, injured people, it really was a terrible situation.' Ignoring his own problems and the danger from exploding ammunition as it detonated in the heat, he stumbled to the cockpit and found Joplin still in his pilot's seat, with both legs broken and tangled up in the rudder pedals. Fish lifted Joplin out, hearing the crunch of broken bones as he pulled him free of his seat. 'I dragged him out,' Fish says, 'but he was a dead weight and I have no idea how I managed it. I'm not a strong fellow at all, so it must have been the adrenalin flowing – in those situations, you just do what you have to, don't you? I can still hear Joppy's bones cracking as I moved him, it really was an awful noise.'

He put Joplin down on the other side of the aircraft to Tilley. 'We shouted to each other,' Tilley says, 'checking up on each other, but there wasn't much to say.'

Fish returned to the aircraft to try to rescue the mid-upper gunner, Bob Yates, and the bomb-aimer, Arthur Walker, but was driven back by the ferocious heat, though he glimpsed their charred bodies through the flames. 'They were clearly dead and I couldn't get to them anyway,' Fish says. 'It was just a mangled wreck and a furnace of flames.' By a cruel irony, Walker's forty-fifth op to complete his two tours had proved to be his last in every sense.

The rear gunner, Flight Sergeant Jim Thompson, had survived the crash, thrown clear when his gun turret was sheared off by the impact, but he had fractured his spine and was stumbling around in agony. The radio operator, 'Cookie' Cooke, had banged his head against his wireless as the plane crashed, fracturing his skull. Although badly dazed, he had also managed to escape, but his smouldering Mae West was burning his wrists and Fish helped him take it off.

The survivors 'were stuck in that freezing field' for about three and a half hours. Tilley remembers:

> There was Joppy over one side. Cookie was wandering round in delirium, because the bang on the head had really brained him, and he'd also burnt his hand. Tommy was wandering around as well, holding his back. And I couldn't go very far on one leg, so I just lay there watching the aircraft burn. That was an awful sight. She had seen us through all those battles and here she was dying. It was, without doubt, the worst time of my life.

The aircraft had crashed in farmland several miles from Ludford Magna, and in the thick fog there was little chance of rescuers finding them quickly, so Basil Fish – though concussed, the only one capable of walking – set off across the field. The fog was so dense that he could barely see where he was placing his feet, let alone what lay around him, and with no stars nor any trace of visible light, he had no idea in which direction to head. He set off blundering across ploughed fields and through hedges, stumbling and falling repeatedly, his head throbbing and blood seeping steadily from his scalp wound. 'It was a terrible journey,' he says. 'I was injured myself and didn't know where the hell I was going. I just knew I had badly wounded friends waiting for help. I was the only one who could do anything and it was the thought of my crew that spurred me on; they would have done the same for me.'

He covered some 3 miles before eventually finding a narrow track leading to an isolated farmhouse. The building was in darkness but he hammered on the door and woke the sleeping farmer. There was no telephone, but the farmer, lighting the way with a Tilley lamp, escorted him to the nearby village, where there was a telephone box. Fish first had to convince the operator that it was an emergency and not an excuse to avoid putting twopence in the slot. She grudgingly put him through to the nearest RAF base, Ludford Magna. He passed on his location, and soon afterwards an ambulance and 'a team of squaddies with stretchers' arrived to begin the rescue.

Fish told them what he could about the length of time he'd been walking and the terrain he had crossed and they began to search for the crash site. Eventually, after repeatedly blowing 'SOS' on whistles as they combed the muddy fields and lanes, they heard a faint answering whistle, and hurrying on through the fog they at last saw the glow of the still-burning aircraft. After initial assessment and treatment of the injured where they lay, Joplin, Tilley and Cooke were taken to the civilian County Infirmary at Louth, while Fish and Thompson – the latter's fractured spine not yet diagnosed – went to the base sick quarters at Ludford Magna, where they were given pain-killers and hot-water bottles, and put to bed. Thompson's spinal fracture was only discovered during a more thorough examination the next day, and he was then transferred to the RAF's specialist injuries unit at Rauceby.

After initial treatment at Louth that night, the RAF reclaimed the others 'in one of their bone-shaking ambulances' and also took them to Rauceby.

Arthur Joplin had been isolated from his friends at Louth and, as no one had told him otherwise, was in the tragic position of believing his full crew had survived the crash. 'On my way, the ambulance crew stopped to check up on me,' he says.

> *The conversation turned to the crash and I said to them, 'At least we all got out.' I remember to this day they turned to look at each other without saying a word. I could see from their faces something was wrong, but they clearly didn't want to tell me what. I pressed them and eventually they told me that both Bob Yates and Arthur Walker had been killed. My whole world collapsed around me – a hole could have opened up and I would gladly have climbed in. I'd spent the last few days thinking how lucky we had all been to escape. Now I had the shattering news that two of my dear friends were dead. As their skipper, I was responsible. It was a terrible time.*

★   ★   ★

Recalling the events from his retirement home in Auckland, the loss and responsibility – even though there was seemingly nothing he could have done – clearly still weigh heavily on Arthur, and his voice falters and dies at the distressing memories.

Frank Tilley's memories of the experience also still loom large. 'I was in the crash ward for a couple of days,' he says with a shudder:

> *seeing sights that you wouldn't want to see again: chaps with no arms, no legs, terribly burnt and dragging themselves around. I was badly injured but this was a real eye-opener. It was the first time I'd really seen, understood, what war meant. It wasn't a nice revelation. One guy's face was burnt off and his bones were sticking through the ends of his fingers – I often wonder what happened to him. But I just wanted to get the hell out of it, and that was actually the end of my flying career. I didn't fly any more because by the time I'd come out of convalescence the war was over.*

While still being treated for their injuries after the crash, Joppy Joplin and Basil Fish were subjected to hostile interviews by senior officers investigating the incident. They concluded that Joplin had been negligent in not adhering to standard regulations governing the approach to the airfield – regulations that were routinely ignored by all pilots, even in clear visibility and with ample fuel in their tanks. Joplin and Fish were both reprimanded and had their logbooks endorsed. 'I thought it was pretty bloody unfair to be given a red endorsement, to be honest,' Fish says. 'It hurt to be held responsible for the crash and of course the deaths.'

'We all thought it quite unfair,' Tilley adds.

I remember after the war, Joppy was very reluctant to come to a reunion we had in New Zealand. He thought everyone might blame him for the deaths of the other crew, but of course no one did; it was just a tragedy of war. It played on my mind for a long time though – it still does, especially not being able to do anything to help. It's there all the time really …
I still think about that. I realise I couldn't do anything, but you can't help wondering.

The night after flying on the Politz op, Ken Gill, the navigator on Barney Gumbley's crew, sat by his bed to write a letter to his parents. 'Last night we were doing a spot of work in the Baltic,' he said. 'That makes our sixth towards our twenty so we are coming along slowly. I'm afraid I shan't be seeing you again this Xmas so I'll have to do without my [Christmas] dinner. I'm hoping to see you in January.'[5]

It would be a truly sad Christmas for Arthur Joplin's crew too, and the New Year would not bring more cheer for 617 Squadron.

★   ★   ★

Tony Iveson was on leave over Christmas 1944 and 'saw all the newsreels on the attack in the Ardennes, but by the time we got back to the squadron, that was all over. I think we knew that the war was on its way out, but it still wasn't over and there were still things to be done. The Russians were approaching the German frontier and we knew it was only a matter of time; it really depended on how long the Germans decided to hold on.'

The first raid of the New Year, on 12 January 1945, saw 617 Squadron tasked with a further attack on naval targets, this time the U-boat pens, German ships and a floating dock at Bergen in Occupied Norway. The Nazi Kriegsmarine had now lost its U-boat bases on the Atlantic coast and in Belgium, but from Bergen the U-boats were still menacing the crucial Arctic convoys to Archangel and Murmansk. There had been snow the previous day and soldiers had to be brought in to clear the runway at Woodhall Spa.

'Usually a [snow]fall means that all crews are put on chipping ice off the aircraft before they can become airborne,' Barney Gumbley wrote, 'but the snow feels very dry and doesn't cling at all. Easily comes off when the motors start up.' However, 12 January dawned very cold but bright and clear. The squadron was again strengthened by Lancasters from 9 Squadron and given an escort of Mustangs flown by Polish pilots. Three Lancasters were detailed to attack the dock and six more the ships, while the remainder of the bomber force would target the U-boat pens.

As well as submarines, the hardened concrete U-boat pens were also used to house E-boats – high-speed torpedo boats that emerged at night to lay mines and sink ships supplying the Allied land forces, then returned to their pens before daybreak. Their speed – faster than any Allied craft – small size and the cover of darkness made them almost invulnerable to attack when at sea, and the pens that housed them were formidable structures, with an 8-foot-thick concrete roof that was impervious to normal bombs ... but then Tallboys were not normal bombs.

After crossing the Lincolnshire coast, the gunners tested their weapons, firing short bursts into the grey waters, then settled down for the long, cold flight across the North Sea. The gunners were huddled in their heated flying suits, but the rest of the crew could use the heat from the engines to warm the cabin and did not need them. The wireless operator controlled the flow of heat and, as Tony Iveson recalled with a chuckle, 'he always complained about "frying, roasting, baking", whilst, further from the heat source, we kept asking him to turn it up.'[6]

The pilots flew in radio silence, leaving most of the flying to 'George' – the automatic pilot – until they approached the Norwegian coast. When they arrived at the target they found it partly hidden by ground haze, which was made worse when the first crew to bomb dropped their Tallboy. It 'threw up a huge cloud of muck and dust' which, in the still, absolutely windless conditions, acted like a dense smokescreen, obscuring the target for the rest. The Mustangs had disappeared to attack the ground defences, leaving the Lancasters circling over the target for half an hour, vulnerable not only to flak but to marauding German fighters. 'If we could not see the target,' Iveson said, 'and with no wind, there was no prospect of getting a clear view, what were we doing there?'[7]

After no fewer than ten unsuccessful runs, Phil Martin radioed Johnny Fauquier and was given permission to switch to the one target he could see, a merchant ship steaming up the bay, but as the warning light illuminated on the bombsight, signalling five seconds to automatic release, there was 'an almighty crash', mixed with a sound like 'hundreds of coins being thrown against the fuselage floor'. The Tallboy dropped harmlessly into the sea, with half of its tail blown away. An 88mm shell had made a direct hit on the bomb-bay, but fortunately for the crew, the Tallboy had absorbed the full force of the explosion without itself detonating and had deflected most of the shrapnel harmlessly away through the bomb-bay doors, which were now riddled with holes. The aircraft's systems were still functioning and Martin was able to return safely to base.

Tony Iveson's crew were also still circling over the target when his rear gunner suddenly shouted, 'We've got fighters, Skipper.' The next moment the Lancaster was being riddled with cannon fire from a Focke-Wulf Fw 190. 'I was conscious of streams of white light going overhead and around me,' Iveson said, 'and then the aircraft juddering. You could hear the shells crashing into the aircraft, and the tailplane and fin had practically disappeared on the port side.' An engine caught fire, normally fatal to an aircraft, especially when the seat of the blaze is close to the fuel tanks, and Iveson gave the precautionary order 'Stand by to abandon aircraft', as he and the flight engineer went through their well-practised drills.

As he did so, 'the Fw 190 did an upward roll right in front of me; I almost heard it, it was so close. We jettisoned the Tallboy but we were in real danger of the fighters finishing us off, and I headed back towards the flak, thinking they might not follow us into the flak. My training just took over; I had practised and prepared for the worst, and when it happened I was ready for it.'

They were still trailing flames and smoke and 'looking rather tattered', and the aircraft was almost unsteerable. Iveson's leg was shaking with the strain of holding the rudder pedal, and the bomb-aimer, Frank Chance, lashed a rope around the pedal to ease the pressure. They had lost the intercom, and when Chance went back to check on the rest of the crew, he found 'three hats on the rear step of the Lancaster and the door open'. The two gunners and the wireless operator had baled out.

All aircrew knew the drill if it became necessary to abandon their aircraft, but very few ever practised it, and one squadron CO regarded it as less important than his men being 'clean and pressed', in case they were taken prisoner. Murray Vagnolini's CO on 61 Squadron had greeted him with the words, 'Welcome to the squadron. Just remember this: while you're here I expect you to have a short haircut, be well washed and shaved, clean underwear and well-pressed trousers. I tell you this because we are all likely to be shot down at some time and that's how I expect you to be received by the enemy.' Vagnolini shakes his head at the memory. 'He didn't want us going over to prison camp with dirty underwear!'[8]

Vagnolini and his crewmates did take one particular precaution before each op, but it had nothing to do with flight safety: 'We'd always check before heading off to see which pubs actually had beer in stock ready for our return,' he says with a grin. 'Sometimes they'd run out or it was rationed. We didn't want to get caught, as our first priority after returning from an op was to head to the pub and down a few beers to celebrate surviving another raid!'

They did sometimes talk about what might happen if they were shot down and they knew of errors that had been made in the heat of a battle. 'If the skipper shouted "Prepare to bale out," we'd heard that a number of crew simply jumped at that point,' he says. 'A couple of friends I knew had been lost over Germany when their aircraft actually made it home! We didn't want to make that mistake, so we practised reacting to the right orders.'

Unlike some of their peers, Tony Iveson's men had practised their drills for abandoning the aircraft, but under the stress of actual combat, just as Vagnolini had noted, three of Iveson's men had mistaken the order. All three landed safely but were captured and became PoWs.

While his colleagues were floating down beneath their parachutes, flight engineer John 'Taff' Phillips went to the back of the aircraft, found some loose ends of the control cables and lashed them together so that Iveson regained some control of the aircraft. However, looking through the astrodome, Phillips could see 'part of the tailplane flapping and part of the rudder had a bloody great hole in it'.

'If the fin had fallen off then we would have gone straight down,' Iveson said, 'but I didn't give any thought to baling out – there was an awful lot of ocean between Bergen and the Shetlands.'

Fauquier was still telling his men to continue orbiting the target, waiting for the smoke and dust to clear, but, said Iveson, 'by that time I was not interested in the target. My only option, flying a crippled aeroplane with a jettisoned bomb, on three engines, with a fire-damaged wing and tailplane shot to bits, lacking a rear gunner and with no top turret anyway, was to get the hell out of it.'

Meanwhile, John Pryor's Lancaster had been attacked by two German fighters, and with one engine silenced, two others badly damaged, and the gun turret disabled, he began to search for a suitable place to crash-land. However, the terrain was so rugged that he decided it was out of the question and instead tried to gain as much height as possible so his crew could bale out. Over the intercom he gave the order: 'Cancel the crash-landing order! Cancel the crash-landing order! Prepare to abandon aircraft! Repeat. Prepare to abandon aircraft!' but he was struggling to get his Lancaster 'G' for George to gain enough height.[9]

One of the German fighters then flew very close alongside and, Pryor says, 'was making signs to me to bale out'. It was a notably humane gesture, for many fighter pilots would simply have continued to fire into the bomber. He flew off but then came back to make a dummy attack, before gesticulating again. Pryor told his navigator: 'I think that bod is trying to tell us that if we don't land he will shoot us down,' but he had 'no intention of presenting the Germans with a Lanc, not even a shot-up one'.

They fired off a couple of red distress flares, hoping to convince the fighter pilot that they were about to bale out, and Pryor managed to coax the Lancaster up to 700 feet, as much as the engines would give him. 'Get ready to leave when I give the word,' he told his crew. 'No hesitation and don't impede the man behind you. Don't rush things when you are out and don't count the usual two before you pull the cord.' Had they completed the usual count, they would have hit the ground before their chutes had fully opened.

He turned the aircraft to seaward and waited until they were approaching a large island before saying 'Go!' One by one they baled out, but determined not to let G for George fall into enemy hands, Pryor waited until he was 'sure she had a watery grave' before following them through the front escape hatch. He landed on the island in snow so deep that he had to feel his way up the cords of his parachute to find his way to the surface of the drift, then buried his harness but kept the parachute silk wrapped around him against the bitter cold.

Pryor walked to a village where he was offered help by members of the Norwegian Resistance, but 'declined their offer for fear of vicious reprisals against the villagers' by the Nazis. Soon afterwards he was captured by the Germans and became a PoW. One of his crew members, who may have been knocked unconscious as he exited the aircraft, failed to deploy his parachute and

died, but the others survived, though, like Pryor, all were captured and became PoWs. Others were not so lucky.

Hit by flak, and with smoke belching from an engine while
German fighters swarmed around his Lancaster, pouring in more
rounds, Ian Ross was forced to ditch in the sea. Two other pilots
from 617 Squadron, Freddy Watts and Jimmy Castagnola, flew to
his assistance, their rear gunners targeting the fighters, and could
see his crew scrambling onto the fuselage. Castagnola dropped
them some Mae West life-jackets and his wireless operator managed to contact a patrolling air–sea rescue aircraft which flew to
the scene and dropped a life-raft.

Watts's rear gunner saw Ross's crew clamber into the raft but,
short of fuel, Watts then had to turn away towards the British
coast, leaving the air–sea rescue aircraft circling the raft. Within
minutes a Focke-Wulf Fw 190 had appeared and, as the unarmed
air–sea rescue aircraft fled, the pilot of the German plane began
strafing the life-raft. All seven crew were killed, and only the
body of the wireless operator, Flying Officer Mowbray Ellwood,
was ever found.

★　　★　　★

With his jury-rigged rudder and control cables, Tony Iveson had
set out to nurse his crippled Lancaster back across the North Sea
to the Shetlands. It was a 340-mile flight and, says Iveson, 'you
cannot measure three hundred and forty individual miles over
the featureless sea, you can only endure it. The relief when at
last I saw the Shetland Islands appearing out of the mist on the
horizon was palpable.'

As they made their approach to Sumburgh airfield, he discovered that the hydraulics had also been shot away and, although
they could still blow down and lock down the undercarriage, it
would have an immediate effect on the Lancaster's already pre-
carious flying performance. He waited until the very last moment
and had just lowered the undercarriage on final approach to the
runway when a Spitfire suddenly cut in front of him, its propeller
motionless and the pilot holding a 'dead stick'. Without any

power, the Spitfire obviously had priority, and Iveson was forced to go around, praying his crippled aircraft would hold up long enough. 'We were all quite detached by then,' he said. 'Either we were going to make it, or we weren't.'

Lining up yet again on finals, he could see the Spitfire being pushed off the far end of the runway. As he landed, the Lancaster swerved off the runway to the left, the tyre of its port under-carriage having been shredded by the crippling cannon fire. They bumped and ground to a halt, got out and, says Iveson, 'about then we started to shake a bit!' His Lancaster never flew again, being classified 'Category E' – unrepairable. It also proved to be Tony Iveson's last operation with 617 Squadron.

'Bergen was talked about for a long time because of the losses,' says one of the survivors of the raid, Colin Cole. 'Two shot down and the Iveson crew shot up. It seemed to affect the squadron for quite a while.'[10]

★　　★　　★

Although the Battle of the Bulge did not end until 25 January, delaying Allied advances in the West, the pace of German defeats in the East was accelerating. The Soviets had brought forward a planned offensive to 3 January, and after liberating Warsaw on the 17th, the Red Army entered East Prussia three days later. After liberating Auschwitz on 27 January, Soviet troops crossed the Oder on the last day of the month, bringing them to within 50 miles of Berlin.

In the air, the pressure on Germany from Bomber Command and the USAAF was unrelenting. On the night of 13/14 February 1945, 800 RAF aircraft launched an attack on Dresden, the ancient capital city of Saxony, close to Germany's eastern borders. The arguments and discussions around that attack still rage to this day, but there is no doubt the raid devastated the city and caused tens of thousands of deaths.

On 1 March 1945 Goebbels wrote in his diary that 'the air war has now turned into a crazy orgy. We are totally defenceless against it. The Reich will be gradually turned into a complete desert.'[11] However, the Luftwaffe was not completely diminished, as its new jet-powered fighters had already begun to appear in the skies over Europe. They were a potent enemy. In one forty-eight-hour period between 20 and 22 February, Bomber Command suffered the loss of 62 aircraft, most to fighters. And on 4 March, German night-fighters followed the Main Force stream back to England, attacking 27 airfields and shooting down some 20 aircraft.[12] Although the war may have been in its final stages, the dangers were still ever-present.

A series of mass daylight raids on Berlin during February carried out by USAAF bombers, and augmented by night-time raids by Bomber Command, reduced much of the city to rubble. On 7 March US ground troops crossed the Rhine for the first time, at Remagen. The question was no longer if, but when, the Nazis would be forced to surrender.

After a long period of raids on naval targets, during February and March 617 Squadron's attacks were chiefly on more traditional targets, including no fewer than five attempts to destroy the Bielefeld railway viaduct in north-west Germany. Four hundred metres long and twenty high, its twenty-eight graceful arches spanned the Johannisbach river at the head of the Obersee Lake. The railway line that ran over the viaduct connected Berlin, Hamburg and north-eastern Germany with the main industrial zones of the Ruhr and was the busiest freight line in Germany, with over 300 trains a day using it.

On 13 March 1945 twenty Lancasters and one Mosquito took off from Woodhall Spa for yet another attempt to flatten it. The scale of the devastation around the viaduct showed how often Allied bombers had previously tried and failed to topple the structure. It had been the target of no fewer than fifty-four previous attacks, but even when the track was struck seventeen times in one raid, the damage was repaired within twenty-four hours, and none of the attacks had had any significant impact on the viaduct itself. Chipped and scarred, it still stood upright, defying everything the Allies could throw at it, though the landscape around it was pockmarked by hundreds of craters, punctuating a sea of torn earth, ruined buildings, uprooted trees and barren, bottomless mud.

In the previous November alone the Americans had dropped over 2,000 high-explosive bombs and 33,000 incendiaries on the viaduct, damaging two spans and a pier, but the Germans had simply placed steel girders over the damaged sections and relaid the track. 'Winter rains turned the crater landscape in[to] a nightmare sea of water-filled holes and deep mud ... The area was alive with unexploded bombs and the bomb disposal workers – the "Suicide Squads" – had a difficult time even reaching the UXBs.'[13]

A Lancaster over Bielefeld viaduct

By mid-March, 'over seven million pounds of high explosive had been dropped on the structure but still it stood, chipped and pock-marked by numerous hits and near misses.'[14] Allied air superiority meant 617 should have a relatively trouble-free flight to the target, and they had attacked the viaduct so often that the navigators could find it without reference to their maps. However, the next attack would have one crucial difference: two of the Lancasters were carrying another brand-new bomb devised by Barnes Wallis.

★   ★   ★

On the morning of 13 March 1945 Wallis took his normal bracing constitutional swim in the frigid waters of the Channel near the Ashley Walk bombing range in Hampshire. Having towelled himself dry and dressed, he took up station behind a blast-wall to witness a live test of his new super-weapon: the monstrous 10-ton bomb that he had christened 'Grand Slam'. It was the largest 'iron bomb' ever built.

Wallis had first proposed the weapon as far back as 1941 in a paper titled 'A Note on a Method of Attacking Axis Powers'. He had suggested that sufficiently large bombs exploding deep underground could destroy targets with the seismic shockwaves transmitted through the earth, rather than the blast effect of high-explosive bombs detonating on impact with the ground. He also designed a huge aircraft, 'The Victory', capable of carrying such bombs, but the Air Ministry rejected that proposal on cost grounds, and since the RAF then had no aircraft capable of carrying such a monstrous weight, his idea of a huge 'earthquake bomb' was also spurned.

Now, three years later, its time had come. He designed the 26-foot-long Grand Slam with an aerodynamic shape and a casing forged from a single piece of perfectly formed chrome molybdenum steel. It took almost a week to cast and machine each casing, and although the exterior was perfectly smooth, the inner surfaces of the casing were often quite rough when cast. When two 617 crews were sent to the English Steel Corporation in Sheffield to boost the morale – and the production rate – of the men manufacturing them, they were shown one casing that was being finished off from inside by 'two little men, smaller than jockeys, with pneumatic hammers'. They shone torches into the casing and whenever they spotted an imperfection, one of them climbed inside and hammered it out. 'He could only stand about twenty or thirty seconds of this and then they hauled him out, his legs shaking.' The insides of the casings were further smoothed with angle-grinders and then varnished inside to remove any risk of premature detonation by friction between the casing and the explosive charge at the moment of impact with the ground.

When a steelworker asked John Langston, one of the visiting aircrew, what he did, he was told, 'I drop the bombs.'

'What do they pay you?'

'Thirteen shillings and sixpence [67½ pence] a day.'

'You're a fool,' the man said. 'I get ten quid a week for doing this!'[15]

The finished casing was strong enough to survive the 10-ton bomb's brutal impact with the ground and allow it to bury itself deep below the surface. A time-delay fuse would then detonate the 9,500 pounds of Torpex explosive (a mixture of TNT, RDX and aluminium powder) it contained, and in theory the shock-waves radiating outwards would create a localised 'earthquake' that would shake any nearby structures so violently that they would inevitably collapse.

Residents in the villages surrounding the Ashley Walk range had not, of course, been told what was being tested that morning, but they had all been advised to keep their windows open to avoid having them shattered by the blast. As the bomb, fitted with an eleven-second delay-fuse, impacted with the ground, Barnes Wallis began silently counting to himself. He had reached nine when the ground erupted. Detonating well below the surface, the visible effects of the blast were perhaps less dramatic than most of the dignitaries assembled to watch the test might have expected, but the inhabitants of the surrounding villages all felt the ground shake from Wallis's 'Earthquake Bomb'.

Unlike the area bombing of the industrial Ruhr, this super-bomb offered the possibility of disrupting or destroying key elements of Germany's industrial infrastructure without causing civilian harm – a particularly important consideration when used against targets in the occupied territories of France, Belgium, the Netherlands, Poland and Scandinavia – and it was hastily pressed into service.

Astonishingly – a new weapon would only be passed for operational use today after being tested for years under every conceivable condition – a handful of hours after Wallis had watched that trial drop on the ranges at Ashley Walk, two aircraft of 617 Squadron were carrying Grand Slams across the Channel, making for the Bielefeld viaduct. The new weapon was so expensive to produce that it was only to be used on the most crucial targets, and those that were not susceptible to attack by more conventional bombs. The viaduct met both of those conditions.

None of 617's existing Lancasters had been powerful enough to lift the massive 10-ton bomb-loads off the runway – the all-up weight of the aircraft when loaded with fuel and the Grand Slam was well over 30 tons – and newer, more powerful B-1 Lancasters, with strengthened undercarriages and more powerful Merlin engines, had been rushed through production to carry them. The bomb-bay doors were removed, the bomb-bay itself faired into the fuselage, and, to save additional weight, each Lancaster carrying a Grand Slam dispensed with its wireless operator and his equipment, the nose turret and mid-upper gun turret and the gunners who operated them, as well as two of the guns and most of the ammunition from the rear turret. As one navigator remarked, they were flying 'virtually naked'.

Even though he was no longer needed as a wireless operator, Colin Cole still flew on some ops 'so I could be with my crew. Just along for the ride really, I didn't want to be left out, though I didn't put those trips in my logbook as we really weren't meant to be doing it.'[16]

Even with all the excess weight removed, as pilot Benny Goodman remarked, 'it was still a major achievement to get a 22,000-pound bomb airborne.' Such was its weight that, after it was loaded aboard the Lancaster, the aircraft's wingtips were 'bent upwards by six to eight inches on either side', said navigator John Langston.[17] It took every inch of runway and every ounce of power that could be coaxed from each Lancaster's four bellowing Merlin engines for the two aircraft carrying Grand Slams, flown by the squadron CO, Group Captain Johnny

Fauquier, and Squadron Leader Charles 'Jock' Calder, to get airborne. Eighteen Lancasters carrying Tallboys accompanied them.

However, yet again their flight across Germany was in vain, for the viaduct was completely obscured by cloud and they had to return home still carrying their Grand Slams. The huge weight of the laden Lancaster would have made it impossible to bring it to a halt before the end of the relatively short runway at Woodhall Spa, and they were diverted to the RAF's emergency landing strip at RAF Carnaby near Bridlington on the Yorkshire coast, where the runway was five times the width of a normal one and 9,000 feet long.

At lunchtime on the following day, 14 March, sixteen Lancasters from 617 Squadron tried again. This time, as Fauquier was testing his aircraft prior to take-off, one of his engines failed. With his crew, he at once scrambled from the aircraft and sprinted across the tarmac to commandeer Calder's Lancaster. However, seeing Fauquier's aircraft still at dispersal with one engine stopped, and the CO running towards him shouting and waving his arms, Calder realised that he was about to lose his own chance of Grand Slam glory, and opened his throttles and taxied away, leaving his furious CO struggling to stand up in the wash from his propellers. As the laden aircraft slowly gained altitude another fourteen Lancasters, which had taken off from Woodhall Spa, joined Calder's aircraft and set course for Bielefeld. There were no other Grand Slams available, and, as before, the remainder of the squadron were carrying Wallis's earlier invention, Tallboy. Regarded as an almost unimaginably devastating weapon just a few months before, it must now have seemed almost puny in comparison with the massive and fearsome Grand Slam.

Flying alongside Calder's aircraft, Sydney Grimes had a ringside view: 'It was the strangest sight I'd ever seen. I could see the wingtips flexing upwards in a bow shape under the strain – the wing was no longer flat. The bomb looked brutal hanging there, a lethal weapon. I thought, If we don't destroy the viaduct with this, then we never will!'[18] 'It looked like a huge, mean bomb,' a member of Calder's crew, Murray Vagnolini, added. 'I did think about the death or destruction it would cause. I'd suffered it all myself during the Blitz so had a good idea what it might be like for the Germans. It must have been pretty awful, but it was war, and what we were doing was a necessary part of that war to defeat the Germans.'[19]

The Germans were well aware of the Bielefeld viaduct's strategic importance, and it was heavily defended. A 20mm flak battery was sited on the hillside above the viaduct and a Wehrmacht Nebel Kompanie – Smoke Company – had stationed its smoke generators along the valley sides. They fired them up when the alarm was given, and by the time the Lancasters had reached the target, the northern end of the viaduct was obscured by smoke, but the southern end was clearly visible as the raiders began their bomb-runs.

The other aircraft dropped eleven Tallboys, but despite a number of near-misses that damaged the viaduct and blocked the train line to the south of it, none of them had demolished the structure. Then Calder began his run. At precisely 4.28 that afternoon, his bomb-aimer, Flight Lieutenant Clifford Crafer, released the very first Grand Slam. The effect on the aircraft, relieved of its 10-ton bomb-load, was immediate and dramatic.

Freed of the massive burden, the aircraft hurtled upwards and Crafer, lying on his stomach in the bomb-aimer's position, was pinned to the floor. When the headlong rise stopped, he 'shot up from the floor and came down with a bang which knocked me breathless'.[20] Sydney Grimes saw the aircraft 'leap upwards', and Murray Vagnolini, who was poised to wind in the retractable arms that had held the bomb when it was dropped, found himself plastered to the floor of the aircraft as it shot up like a cork from a bottle. 'When an aircraft dropped a bomb that size it leapt about two to three hundred feet into the air, but this was something they didn't tell us,' he recalled with a rueful smile.[21]

Dropped from just under 12,000 feet, the Grand Slam cut a graceful curve on its thirty-five-second fall through the air, as the spin imparted by its vanes – like the Tallboy's, offset at an angle of 5 degrees – kept it tracking unerringly towards the target, the central span of the viaduct.

Keen eyes in the circling Lancasters saw the 'squirt of mud as it speared into the marsh' about 20 feet from the stone piers. Impacting at near-supersonic speed, it drilled down deep into the muddy earth alongside it. Eleven seconds later, the time-delay fuse triggered a massive explosion, creating a crater 60 feet deep and 200 feet wide, blasting debris hundreds of feet into the air and sending up a towering column of dust and smoke.

'It exploded with a fantastic flash which was at once smothered by a high column of smoke,' said Clifford Crafer, whose aircraft was battered by the force of the explosion, which he described as 'worse than from the severest flak'.

Jock Calder described how their 18-ton Lancaster was tossed around by the blast as if it were a paper bag in a gale, later saying that 'it felt as though someone had hit me severely in the back. I didn't expect the kick quite so soon.' The shockwave from the explosion swept his aircraft 'well over 500 feet' higher in the sky.[22]

When the smoke and dust settled, they could see at once that their mission had been completely successful. The 'Earthquake Effect' that Barnes Wallis had theorised had shaken the viaduct to its core, and seven arches, incorporating a 200-foot stretch of the northern span and 260 feet of the southern one, had been completely obliterated.[23] One watcher said that the entire viaduct seemed to be lifted into the air before crumpling in a heap. A single 10-ton bomb had achieved what a total of 3,500 tons on previous raids had failed to do.

Anneliese Möller, a member of the first-aid unit in a nearby bomb shelter, recalled that 'the ground vibrated like an earthquake' when the bomb detonated, 'and people realised at once the viaduct was hit. As we [arrived at the site] we found thirteen [dead] anti-aircraft auxiliaries on the ground. A horrifying image.' Erna Bitter, a schoolchild at the time, saw the explosion from the windows of her school and watched as 'the dead were brought to the funeral chapel on handcarts. There were many carts.'

Twelve-year-old Helmut Brockmann, watching from his garden only 2 miles away, remembered clearly seeing the 10-ton bomb fall towards the viaduct. 'She plummeted from the sky like a telephone pole,' he says. 'I had never seen anything like it before. Then I felt the blast, my trousers were fluttering in the wind, and the earth vibrated. Women in the shelter began to scream. When I looked again, the viaduct had gone.'[24]

A Grand Slam is released over the Arnsberg viaduct

No doubt to the continued fury of his Commanding Officer, Calder's reward for being the first man to drop a 'Ten-Ton Tess' was a Bar to his DSO (he also had a DFC).[25] However, Johnny Fauquier's chance to drop his own Grand Slam came in a similar attack on the Arnsberg viaduct the next day. Poor weather disrupted that raid, and the viaduct was undamaged, but four days later 617 Squadron returned to finish the job. Fauquier's bomb was released early and undershot, but Phil Martin's bomb-aimer, Don Day, made no mistake. A squadron leader from a Main Force bomber squadron, along for the ride as an observer, was peering over Day's shoulder as he released the bomb from 12,000 feet and scored a bullseye on the western end of the viaduct. A few seconds after it detonated, 'we were thrown upwards and almost went upside-down from the concussion because of our bomb hit. About four Tallboys also hit and exploded within, I would say, about two seconds, and we just ended up a tangled heap over the bomb sight!'[26] When the smoke cleared, two arches of the viaduct had collapsed.

There was also a terrible human cost from the raid. According to Benny Goodman, who paid a visit to Arnsberg after the war, the Germans had put 'a whole school and its teachers' under the arches of the bridge, and when it was bombed they were all suffocated.[27]

The Bielefeld op proved to be Sydney Grimes's last. As more of the 617 Squadron Lancasters were converted to carry Grand Slams, the mid-upper gunners and wireless operators found themselves redundant. Grimes was:

very resentful that I'd only got seventeen ops in my second tour, when we should have done twenty. My crew would complete the full tour and I was being left out. I wasn't happy about that at all. I really wanted to finish with them, there was a sense of letting them down, not doing 'your bit', perhaps of taking the easy option out of the war. I certainly didn't want to be accused of that! I asked my Flight Commander if I could go on ops anyway, and just stand in the astrodome and look out for flak and fighters, anything to get the last three in! I tried my best to persuade him but he showed me a signal from 5 Group telling the squadron that no spare bods would be taken on any ops. That was that.[28]

★　★　★

On 21 March 1945, Vera Gill began a letter to her beloved husband Ken, the navigator on Grimes's crew. 'My darling hubby,' she wrote. 'Isn't it a lovely day for the 1st of Spring, I wish we could be together to enjoy it. I wonder if you have been [flying] this morning; it's just been on the 1 o'clock news. Of course I know there are plenty more Lancasters besides yours but I just wondered.' She gives him news of their baby son Derek and goes on to thank him for the oranges he had sent and the photograph which had arrived a few days before. It seemed Ken had not been happy with the results of the photo of the pair, as Vera wrote, 'I think you must have stood nearer the camera! Even if you were as fat as that I would still love you the same.' The letter ends abruptly there. Presumably there was a knock at the front door.

Twenty-two-year-old Yorkshireman Ken Gill was a veteran of forty-six operations and had already been awarded a Distinguished Flying Cross after his first tour, though he had not yet had a chance to attend a presentation ceremony. He clearly adored his young wife, starting his letters 'Darling Heart' and signing off with so many kisses they run off the end of the page. In one he responds to her concerns about the daily dangers he faced, writing, 'Yes Sweetheart, we do "daylight ops" but I've changed my mind about them now; you don't feel half so scared when you see what's going on.' He also wrote of his huge affection for his infant son. 'So Derek is a tinker? Oh well, I suppose all lads are but he'll calm down. Keep well and smiling, your loving hubby, Ken. xxx'

As Vera had suspected, her husband was airborne that bright spring day. And it was a daylight operation. While the Main Force targeted Bremen, 617 Squadron, including Grimes's old crew piloted by Barney Gumbley, were on a raid against the Arbergen railway bridge a few miles to the south of the city. As Gumbley and his flight engineer, Eddie Barnett, 'a tall, fair-haired, good-looking Londoner', were putting on their flying kit in the locker room, the rear gunner of another crew, Bob Barry, asked Gumbley, who by this time 'had amassed a vast total of missions': 'When are you going to give it [up] Barney?'[29]

'Oh, pretty soon,' Gumbley said. 'I've just about had enough.'

Although he'd been stood down from ops when the wireless operators and mid-upper gunners were dispensed with to save weight, Sydney Grimes was still 'seeing my crew and chatting to them. I used to go up to the Flights with them in the morning, just seeing what was going on, still trying to be part of it really.' He and Gumbley had already made plans for what they would do when the war was over. Grimes recalls:

Barney's father had a chain of mobile cinemas which went around to all the little towns in New Zealand showing movies. He wanted to extend the business when Barney came home and asked if some of the crew might like to move out there and be part of the bigger business. I was very tempted. It seemed like a great opportunity. It was unusual to talk about the future, but we had a sense the war might soon be over so it started to come up in conversation. But I didn't give too much thought to it because I didn't want to tempt fate. I was married, Iris was taking her final exams, we all needed to think about what we were going to do – but it was all in the future. I didn't dwell on it. That would have been bad luck whilst we were still fighting the Germans.

The night before the Arbergen op, Grimes had been in the Mess with his crew as usual. 'It was just a normal evening,' he says. 'They'd been told they were on ops next day so had an early call at five a.m., but I never thought about getting up to see them off. I rather thought that might be bad luck, so just told them I'd see them the next day.'

The raid was successful. A German civilian who witnessed the attack described one of the falling bombs as 'like the outlines of a diving whale'. Relieved of its massive burden, he watched the aircraft as it 'shot up perhaps a hundred and fifty to two hundred metres, like a fast elevator on its way to the top. Then there was a gong-like sound, very strange, metallic,' as the bomb detonated.[30] Two arches of the bridge were destroyed by the effects of numerous near-misses from Tallboys and Grand Slams.

The human impact of such raids on civilians was vividly demonstrated when an elderly lady's house near the bridge was razed to the ground by the bombs. There was a small bomb-shelter behind the house, and the lady's son hurried there but found it completely buried by rubble. He began tearing at the debris with his bare hands and discovered the body of his mother. Neighbours 'took him aside in order to save him from the terrible sight'. Several of her neighbours also died in the shelter. The pressure wave from the blast had blown the woman's cast-iron sewing machine into an apple tree 50 yards from the house. It was the only one of his mother's possessions that he was able to retrieve.[31]

However, there were heavy casualties among the attackers as well. The Lancaster crews had to fly through a dense flak barrage and could also see the new German Messerschmitt Me 262 jet fighters – the world's first operational jets – 'easily dodge the Spitfires, Mustangs and Lightnings which chased them all over the sky while the jets picked off Lancasters at will'.

As they began their bomb-run, rear gunner Bob Barry saw the neighbouring Lancaster, piloted by Barney Gumbley, hit in the starboard wing, either by fire from a fighter or by flak. More than sixty years later, that moment was still seared into his memory:

> It started a death dive and then spiralled a few thousand feet before it exploded. There was no chance for anyone to bale out. I still have a crystal-clear picture in my mind of Eddie Barnett's white face staring helplessly out of the cockpit's starboard window. He appeared to be looking directly at me, but I suppose he was staring in horror at the space where the wing used to be. Like so many others, Barney just didn't make up his mind soon enough; nor did the crew.[32]

The burning Lancaster crashed in a field near the village of Okel, 7 miles from the Arbergen bridge, and its bomb-load detonated moments later. The violent explosion 'shook the entire countryside', reduced the aircraft to 'minute fragments' and left a crater 60 feet deep and 100 feet across.[33] All five crew died instantly.

Sixty years later Gumbley's nephew John visited the crash site and was directed to a nearby house. Frau Else Lippmann answered the door and, astonishingly, even though all those years had passed, 'knew immediately what we were enquiring about when a plane crash was mentioned'. Ten years old at the time, she and a friend described seeing the bomber flying 'unusually low' compared with the aircraft they normally saw passing high overhead. They then saw a lone German fighter plane follow it and fire a short burst into it. Smoke trailed from the bomber and they were scared that it would hit the village, but then it suddenly went down in a field, 'exploding and shaking the whole village. [Else] ran out onto the field to look and was followed by her mother, scolding her for being so naughty. Then a policeman came and told the children not to go out onto the field as it was too dangerous.'[34]

Waiting for his friends back at Woodhall Spa, Sydney Grimes watched the other Lancasters return from the raid, and his anxiety turned to dread as the minutes, lengthening into hours, dragged by with no sign of them. The agonising suspense finally ended when one of the returning crewmen broke the news to him. His crew had perished. 'These were my friends who I'd fought alongside,' Grimes says, 'and been through so much with. If I'd had my way, I'd have been on that aircraft too.'

'It was such a shock when his crew were killed,' his wife Iris remembers. 'I think he kept it from me for a while so as not to bother me, but you just hoped they might have survived somehow and be prisoners … but of course they hadn't. It wasn't to be, and that was so sad.'

There was even worse to come. The crew's flight engineer, Eddie Barnett, was 'a fellow "Townie" – a Southend-on-Sea man', Grimes says. 'His mother received the telegram "missing, presumed killed" and she knew my own mother lived close by, so she got on the bus to our village and knocked at my mother's door, saying what terrible news it all was that we were all dead. My mother didn't know that I was no longer on ops, so they were crying in each other's arms, talking about the loss of their sons.'[35]

In the chaos and confusion, Grimes's mother did not call his wife, Iris, to check because 'she didn't want to intrude' on her grief. 'She didn't want to trouble me,' Iris says, 'so she went to see my mother and told her what she'd heard. It was a terrible time for parents. I think it was worse for the mothers. We were young, eager, perhaps naive, but they knew the realities and always feared the worst.' When Grimes's brother came home on leave from the Navy that weekend, he wondered why neither Iris nor the squadron had contacted his mother, and rang the Mess at Petwood to ask for details of his brother's death. 'Oh no, he's still here,' someone said and called Sydney to the phone.

There were no funerals for his missing crewmates – there were no bodies to bury, they had just gone. 'It was a big loss, though it wasn't an unusual thing to lose your friends,' Grimes says.

The old sweats used to advise us not to make close friends at all, then you couldn't be shocked by the loss. To survive, you all needed the basic skills and you really needed to get on well and work well together, but you also needed a good bit of luck. I guess the day Barney was shot down, the luck ran out, but a few feet difference in the sky, a slightly different speed, and they might have survived. It was just six weeks before the end of the war.

That I could and should have been on that flight with them was a terrible thought at the time, but I do think someone 'up there' must have been looking after me, telling me I had a future.

The loss of his entire crew increased the stress on Grimes and, he says, 'though my nerves never went, after Barney and the rest were killed, Iris said she could see a twitch developing in my cheek. You could see if someone was on the edge – they developed the "twitch". But mine didn't get too bad, I suppose because the war was nigh on over and I'd not be going back on ops.'

★   ★   ★

At her home in Leeds, if Vera Gill was distracted by a knock at the door as she was writing to her husband Ken, Gumbley's navigator, one can only imagine the shock and instant understanding she must have experienced at seeing the telegram boy holding out a flimsy sheet of paper. The telegram was just one of many tens of thousands delivered to the relatives of the men of Bomber Command during the course of the war. It would begin, 'regret to inform you ...'

The letter she was writing to her husband was never finished.

CHAPTER 15

The Final Days

Despite the human cost, the first Grand Slam operations had been a startling success, and they marked the start of a new phase for the squadron, which went on to attack several more major railway viaducts across Germany, including Bremen and Nienburg, but it was another kind of industrial target that gave the most vivid demonstration of the Grand Slam's war-winning potential.

Military Intelligence had reported that 10,000 slave labourers were building a vast underground U-boat factory, code-named 'Valentin', near the Bremen suburb of Farge, on the river Weser, 40 kilometres upstream from Bremerhaven. Many of the labourers – German criminals and political prisoners, and French, Polish and Russian prisoners of war – were housed in a nearby concentration camp, and construction continued round the clock. Estimates of the death rate among the slave labourers range as high as 50 per cent. Life expectancy among the 'iron detachments' – *Eisenkommandos* – responsible for erecting the iron and steel girders was so low that one French survivor, Raymond Portefaix, gave them a different name: 'suicide squads'.[1] The factory they were building was to be used to construct a fleet of Type XXI super-submarines that were far faster, quieter and more deadly than any of the existing U-boats that were still sinking millions of tons of shipping.

The gateway in the factory's western wall giving access to the river could be closed by a sliding bomb-proof door, and the factory itself was shielded by a concrete roof that was 14 feet thick at its western end and 23 feet thick on the eastern side – the biggest concrete structure in the world. A section of the factory could even be flooded to fully submerge each huge submarine for testing. Defended by dense anti-aircraft batteries, it was designed to withstand anything the Allies could throw at it, while beneath that protective shield, the factory's slave labourers could produce a minimum of three super-submarines a week.

Harry Callan was one of thirty-two Irish-born British Merchant seamen incarcerated in the Farge camp and used as slave labourers in the building of Valentin. His daughter-in-law described his experience: 'They were forced to work for over a year on the building on starvation rations. Body weight was more than halved in that time. Harry went from twelve stone to barely six stone in weight. Bunker Valentin was HELL. He is alive today due to the kindness and courage of some German families from the Farge and surrounding villages.'[2]

Allied commanders knew that the factory was under construction, but wanting the waste of Nazi resources to be as great as possible, they had held off from attacking it until it was almost finished. Only on 27 March 1945, when the factory was 90 per cent complete, with most of the machinery installed, and work beginning to pour the concrete for the roof that would have rendered it virtually invulnerable to air attack, was 617 Squadron ordered into action.

When the target was revealed at the briefing, the members of the most experienced and 'probably the most "flak-happy"' crews on the squadron were perturbed to discover that it was at Bremen, 'a target to which we had been before in daylight and which had us fairly scared because our welcome had not been very pleasant on that occasion.'[3]

Their nerves caused the smokers among them to get through even more cigarettes than usual at the briefing, but as one of them stubbed out a cigarette, he failed to notice that a squadron humourist – possibly their CO, Johnny Fauquier himself, whose default stern expression hid a dry sense of humour – had connected the two ashtrays on their table with a line of cordite. The next moment there was 'an explosion and an enormous puff of smoke! Seven flak-happy crewmen jumped to their feet and clutched each other like frightened children. Fauquier gave the ghost of a smile and said, very quietly, "Okay fellers, now let's get on with the briefing!"[4]

Johnny Fauquier with a Grand Slam

Twenty Lancasters of 617 Squadron, six carrying Tallboys and fourteen with Grand Slams fitted with time-delay fuses, were to target the U-boat pens, while ninety-five Lancasters from other squadrons in 5 Group made a simultaneous attack on the nearby oil storage depot. Two of 617's pilots had to abort and turn back. One of them, Benny Goodman, had engine trouble from take-off and knew he was not likely to get to the target, but he had pressed ahead in the hope of sorting out the problem. He remembered:

The last thing I wanted to do was to have to 'boomerang', but we struggled to climb and just kept falling back, and could never have reached bombing height or got to the target with the rest of the squadron. It was always a very tough decision to turn back, but particularly so on 617 Squadron – it just wasn't done. You had to have a damn good reason. Eventually I had to make the decision to return to base, but I still think about that decision today. Seventy years on, it still runs through my mind. Even though I know I made the right decision, it's a blemish on my record I can't erase.

As Benny sips coffee in the lounge of a Bracknell hotel near his home, anguish is etched on his face as he struggles – needlessly – to explain his decision:

If I'd pressed on, we would have been over the target at perhaps five thou-sand feet and on our own, so we could easily have been shot down and lost a crew … and then I would have been criticised for bad judgement! There was nothing I could do but I felt so dispirited and my morale really took a hit. It was all about the squadron's reputation and personal pride and ethos – all self-inflicted of course, but the pressure was immense. You just don't *turn back!*

The remaining crews reached the target and dropped eighteen bombs – eleven of them Grand Slams – from 18,000 feet. Sixteen fell among the buildings surrounding the complex, completely destroying them. The other two, both Grand Slams, hit the target dead centre at supersonic speed and bored down through the half-set concrete roof to bury themselves deep inside the struc-ture. For some minutes nothing else happened, and guards and slave labourers were just emerging from their air-raid shelters when the time-delay fuses triggered the bombs. The blasts shook the complex like an earthquake, killing 300 men inside it and throwing those in the surrounding area off their feet. Harry Callan, working as a slave labourer, remembers seeing the bombs drop on the bunker with mixed feelings. He was excited that the Allies had arrived, but feared for the German families who had befriended the brutalised construction gangs. His lasting thoughts were for the many who had died building the 'monster' and that their last work was being destroyed.[5]

When the smoke and dust cleared, save for the twin holes in the roof, the structure looked unaltered from the outside, but the interior was so badly damaged – a jumble of thousands of tons of shattered concrete and twisted steel – that the site had to be abandoned. The tottering Nazi regime's prodigious investment of money, manpower and scarce resources – a last throw of the dice in an effort to win the war at sea – had come to nothing.

Soon after the end of the war, Murray Vagnolini went back to the site with Jock Calder. 'I've a photo of me standing at one of these holes, an enormous crater,' Vagnolini says. Down below the hole, they could see the wreckage of 'about eight submarines' that had been thrown around by the bomb blasts and smashed together like children's toys.[6]

The remains of the massive structure have proved indestructible to this day. Just after the war it was used as a test-bed in 'Project Ruby' – a joint British and American evaluation of penetration bombs that saw well over a hundred bombs dropped on it without significant effect – and after being used for fifty years as a storage depot by the German Navy, it is now a museum and memorial: Denkort Bunker Valentin. Dark, dank and haunting, it is a stark memorial to some 4,000 souls who died during its construction.

A fortnight later, on 9 April 1945, the squadron was tasked with an attack on the U-boat pens at Hamburg. Benny Goodman, still smarting over his aborted mission on the previous op, was one of seventeen crews who took part. This time he reached the target without problems, but once more his luck was out. 'Our bomb hung up and simply didn't come off,' he says. 'We couldn't close the bomb-doors, so we had to turn for home with this damn thing still on board. Suddenly it came free.' It fell not on the submarine pens or docks but in a nearby residential area, and its detonation caused carnage among the civilian inhabitants. 'I felt really bad about hitting them,' he says. 'They were women and kids, and it really made me think about what we were doing, but there was nothing we could have done – the bomb just fell off.'

Although all crews returned safely from the raid, Goodman's had a very lucky escape when an Me 262 jet fighter ranged up alongside them. Much faster and better armed than any British fighter, it was a lethal adversary, with well over 500 claimed 'kills' against Allied aircraft by the end of the war. Goodman says:

My flight engineer, who sat next to me in the jump seat – over Germany, two pairs of eyes were obviously much better than one! – nudged me and sort of nodded his head. I looked down at the instruments and I could see nothing wrong, so I went on flying. So he nudged me rather harder and moved his head more. I looked out of the starboard window and I was horrified to see the latest German fighter, an Me 262, in formation with us.

The first thing that came into my head was the fact that we'd accidentally killed people in Hamburg and this was the payback. I really thought this was it. It didn't look like any machine I'd seen before. I'd never seen the enemy close up like this, especially not in the latest jet fighter. It was all a little surreal. It was on our wing tip and there was simply nothing I could do. I stared over at him but that was it; there was no cheery wave between foes, there was no acknowledgement of each other. I told the crew there was an Me 262 on our wing and they just laughed in disbelief. Then the nav went into the astrodome and said, 'Christ, it's there!' Suddenly it just banked away. All I could think was that he must have used all his ammunition shooting up other aircraft. Otherwise we would have been shot down. Another lucky escape!

★   ★   ★

The war in Europe was now entering its endgame. The Battle of Berlin began on 21 April 1945, the first meeting between US and Soviet forces took place on 25 April, and within two days Berlin was completely encircled by Allied troops. On 25 April, with Soviet artillery already pounding the heart of Berlin, Soviet troops fighting their way into the city and Adolf Hitler's tottering regime about to fall, 617 Squadron launched their final op of the war: an attack on the Berghof – 'The Eagle's Nest' – Adolf Hitler's heavily fortified last redoubt, high in the Bavarian Alps.

617 Squadron would the attack, dropping Grand Slams and Tallboys and marking the target for the 400 Main Force aircraft that would follow, obliterating not only Hitler's eyrie but the SS barracks and weapon stores clustered around Berchtesgaden. There were reports from Military Intelligence that Hitler was making his last stand there, and the aircrews were not expecting a cakewalk, believing that it would be very heavily defended. Some crews were more nervous about the op than others, not

least David Ware's, flying with 635 Squadron. They had already flown forty-four ops, Berchtesgaden would be their forty-fifth, completing their tour of duty, and 'we didn't want to get shot down on our last trip'.[7] Benny Goodman remembers that 'there was no sense that this was a final op – it was just another target. We were still expecting a lot of flak defending the area!'

The return flight would take seven hours fifteen minutes. They took off at dawn, flew right down the length of France, turned to the east at the Swiss border and flew on as if heading for Munich, before swinging onto a course for Berchtesgaden. Such was the Allies' crushing air superiority by this stage of the war that there were almost no German fighters left to defend the target. 'We were bristling with fighters,' David Ware recalled, 'total air superiority.'[8] However, one Polish Mustang pilot with 317 Squadron described the anti-aircraft fire as 'terrible' and the sky 'absolutely black' with flak bursts. 'I remember one Lancaster went over the target and missed it for some reason or another,' he said. 'And he turned round and went again a second time. We were trembling in view of the very heavy anti-aircraft fire, but there you are, British phlegm! And on the second round he dropped his bomb and joined the rest on his way back to England.'[9]

Ware had been the first to drop a marker flare, 'Not that you really needed them,' he said, 'you could see Berchtesgaden on the mountain very clearly. We had a great run in and marking was very accurate.' Although Ware had a clear view, intermittent cloud, mist and snow cover on the ground were severe obstacles for 617's crews, making it particularly difficult to identify Hitler's personal complex, which in the event escaped serious damage. However, the SS barracks and fortifications were pulverised by 617's Tallboys and the subsequent pounding by Main Force's bombs.

After the raid was over it was revealed that Hitler was still in his Berlin bunker and had not been at Berchtesgaden anyway, but the attack on his beloved mountain eyrie was of symbolic importance, and gave huge psychological satisfaction to the aircrews. After five and a half years, they were finally taking the war to Hitler's own doorstep, though as Tony Iveson pointed out, 'On the one hand, it was a satisfying two-finger gesture from RAF Bomber Command to the man who had started the war. On the other hand, two Lancasters were lost while making that gesture.'[10]

★   ★   ★

Just a few days later, on 30 April 1945, Hitler committed suicide in his Berlin bunker, and on 8 May – VE Day – the Allies formally accepted Germany's unconditional surrender. The war in Europe was over at last. But the men of Bomber Command had continued to fight and die until the final hours of the war. On the night of 2/3 May 1945, they attacked German ships assembling at Kiel, presumed to be transporting troops to Norway to continue the fight there. Three aircraft were lost and fifteen airmen killed. They were the final Bomber Command casualties of the war, for just a few hours later German officers met with General Montgomery on Lüneburg Heath to sign the surrender. In the eleven months since D-Day, the British Army had lost nearly 40,000 men during the invasion campaign. During the same period, Bomber Command had lost 2,128 aircraft, with some 10,000 airmen killed.[11]

★   ★   ★

VE Day is imprinted on the minds of those who fought and survived. Les Munro was on leave in London and, he says, 'the news just seemed to filter through the streets. People started to flood out – a mass of celebrating men and women, cheering, shouting, kissing each other. I think I kissed a few English girls that day! It was a totally amazing, very memorable day.'[12]

'There was a huge celebration,' Basil Fish says. 'I could hardly believe I'd made it through, to be honest. In many ways, I had no right at all to be alive. But there you go. I certainly partook of a few drinks to celebrate my survival!' People were dancing in the streets of every town and city in Britain, enough alcohol was consumed to refloat the *Tirpitz*, and the seeds for at least part of the post-war 'baby boom' were sown that night. But for others VE Day was no time for celebration. Those grieving for lost loved ones wanted no further reminders of the war, and no street party could ease the pain they felt in their hearts.

Basil Fish and Lofty Hebbard at a reunion

For Benny Goodman too, VE Day was very low-key. 'The news was passed around fairly quickly. I remember the Station Commander addressing us to say that we'd be returning to a "peacetime" RAF: five-day weeks, dining-in nights once a month, parades, sports afternoons. Everything back to normal, just like that! Of course it didn't happen that way.'

John Bell was on leave at home in Epsom, Surrey, with his wife Florence and their four-month-old daughter when VE Day was announced. He 'borrowed' three huge signal rockets from the Air Traffic Control tower and ran straight out to the back garden and set them off:

It was a family occasion, with Florence and our daughter, who had been born on 29 December 1944. We were overjoyed and relieved and I began to realise how lucky I was – I'd completed fifty ops and made it through. The war was over. When it had all started, we had no concept that we would be at war for over five years – I thought it would be over in months. I had been a schoolboy back then, but now I was a man. I had responsibilities and life had to

go on but I had no idea what was ahead of me. So, amidst the joy and relief,
I certainly was thinking, What next?

A week later, on 15 May 1944, some of the 617 Squadron ground crews were given a so-called 'Cook's Tour' of some of the squadron's key targets during the war, to show them the destruction they had helped to inflict on the enemy. The round trip took in Hamburg, Brunswick, Hanover, Bielefeld, the Dortmund–Ems Canal, the Möhne dam, Cologne, the Ruhr Valley, Wesel and Ijmuiden, before returning to Woodhall Spa.[13] They saw the utter devastation caused by Main Force raids on German cities and the more precise destruction inflicted by 617 Squadron. The giant craters left by Tallboys and Grand Slams were still clearly visible. Some are still evident today.

By the end of the war, 617 Squadron had flown some 100 special ops, often against crucial targets only they had the expertise and bombing accuracy to attack, and the successes they achieved are reflected in the more than 160 gallantry awards received by members of the squadron. But those successes came at a heavy price. Thirty-four of the squadron's Lancasters were shot down and almost 200 men had been killed.

In April 1943, on the eve of the Dams raid, 617 Squadron had a strength of 58 officers and 481 NCOs. Of the 133 men who took part in the Dams raid, 56 died on that raid alone and another 45 perished in subsequent ops with 617 or other squadrons; a total of 101 out of those original 133 Dambusters did not survive to the end of the war, a fatality rate of over 75 per cent. Such a rate of loss even exceeds the horrific figure for Bomber Command as a whole, which out of 116,000 men, lost 55,573 killed, 8,403 wounded and 9,835 as PoWs. On one single night – the disastrous raid on Nuremberg – Bomber Command's losses exceeded Fighter Command's during the whole of the Battle of Britain.[14] Arguably no military force in history had ever suffered losses on such an overwhelming scale.

On 22 May 1945, Barnes Wallis was flown out to Bielefeld, and in unseasonably bitter wind and rain he inspected the devastation his Grand Slam bomb had wrought on the previously indestructible viaduct. It was to be almost forty years before the last of the debris was cleared from the site. From the 'bouncing bomb' – the Upkeep – through the Tallboy to the Grand Slam, Wallis had been inextricably linked with the story of 617 Squadron, and had earned the right to be talked of in the same breath as its heroes, men like Gibson, Cheshire and Tait. Together they had found ways to revolutionise the air war, striking a series of devastating blows against Nazi Germany that not only arguably shortened the war, but also played a huge part in lifting the morale of the nation and the Allies. After the Flood had come the Victory.

CHAPTER 16

Counting the Cost

The end came quickly for 617 Squadron. In June 1945, Frank Tilley was released from hospital, where he had been recuperating from the broken leg he suffered in the disastrous crash that claimed the lives of two of his crew after the Politz raid. He was sent back to Coningsby, but by then the squadron had already been transferred to RAF Waddington and Tilley was left:

> *feeling like a loose end, not tied up. My kit had gone from my hut, cleared away by the Committee of Adjustment – I found it in a hangar with all the other belongings of the dead and missing, piled up in a corner. Lots of things were gone: bits of aircrew kit, my camera. Of course, when I returned my kit to stores I had to pay for the items stolen when I'd been in hospital. I was very bitter about that.*
>
> *I hated that time after the war. The squadron had moved on and I was simply forgotten. After being messed about from pillar to post I ended up as a clerk at Catterick! I couldn't get out of the RAF until February 1947. It was a very sad end to my RAF career; I felt rather let down.*[1]

That sense of anticlimax also extended to many of his former comrades. At the end of the war, the huge contribution that Bomber Command had made to the overall Allied victory had been both understood and appreciated by the vast majority of the British public, but within weeks that view began to change. Voices were soon raised, questioning the policy of 'area bombing' and the targeting of cities and civilians, with the attack on Dresden in particular singled out for fierce criticism. It was even described as 'a holocaust', which those who had seen the Nazi concentration camps and gas chambers rightly found an odious comparison.

Winston Churchill had been among the first to denigrate Bomber Command, albeit by omission. In the dying days of the war in Europe, in March 1945, Sir Archibald Sinclair, the Secretary of State for Air, had eulogised Bomber Command for 'hampering and enfeebling the power of Germany in every element. Allied bombing is on such a colossal scale that Dr Goebbels has had to admit that "it can hardly be borne". The swelling crescendo of destruction is engulfing oil plants, tank factories and the communications of the German armies on every front as the Allied armies surge forward into Germany.'[2]

Yet, astonishingly, when Churchill made his victory broadcast to the nation, Arthur 'Bomber' Harris listened with 'mounting incredulity'[3] as Churchill's recital of the key contributions to the victory and the pivotal events of the war, great and small, contained not a single word about Bomber Command.

That indefatigable warrior Churchill now had a new campaign to pursue. The Soviet Union, until recently a close ally, was now recast as the enemy. West Germany, the hated and recently defeated adversary, had become a new front-line friend in the unfolding confrontation with the Soviet Bloc.

In that altered landscape, the Second World War bombing campaign against Germany was now an embarrassment, to be ignored or even deprecated. No special campaign medal was struck for Bomber Command air and ground crews, no memorial erected to the more than 55,000 dead pilots, navigators, flight engineers, wireless operators, bomb-aimers and gunners. As the years passed, generations growing up who had never known war came to deplore and even vilify the role that Bomber Command and its commander had played.

Only in 2012, when a memorial to the men of Bomber Command was finally unveiled in Green Park in London, can the tide be said to have fully turned, though even then, protesters defaced the memorial and Bomber Harris's statue the following year. Despite such incidents, there has been a growing recognition of the vital role that Bomber Command played in the Allied victory, but for the men who carried out that bombing campaign, not least those of 617 Squadron, a bitter taste remains, especially at the subsequent vandalism of the memorial. 'Why deface the Bomber Command memorial?' Benny Goodman says, his voice reflecting his bafflement. 'Why did they pelt Harris's statue with paint? It really was terrible. I think some people forget they are free because of the sacrifice that was made seventy years ago.'

'I resent the fact that it took so long for Bomber Command to be recognised,' Frank Tilley adds. 'Not me and 617 Squadron, but the guys on Main Force, what they went through, the terrible losses they endured.'

'We were treated as heroes during the war, and I've been lucky to have a wonderful family, which has enabled me to keep a sense of perspective about what is important in life,' Fred Sutherland says, 'but the way Bomber Command was denigrated after the war appalled me.'

None of the veterans of 617 the author spoke to gloried in what they had done. They were – and are – only too well aware of the human cost of bombing. Sydney Grimes had been:

religious before the war, but my faith diminished when I saw the effects of the bombing. After the war, in late 1945, I was an adjutant based near Hamburg and I was shocked by the sight of the city. Whole areas were a wasteland – no bricks, no buildings, nothing, just the entrances to cellars; the houses didn't exist, but people still lived in their cellars. That really affected me. I understood what we had done in Bomber Command. I still thought it was necessary and the only way to defeat the Nazis, but I now saw war in its entirety and how those on the receiving end were affected. I had always understood the bomber, but now I was understanding the bombed. Those sights still live in my memory.

He also saw for himself the terrible reality of the Nazi regime:

I was stationed near Belsen. I went past the camp and there wasn't a sound. It was totally silent, no birds singing or trees rustling. It had an evil feel about

it; here was the face of Nazism, the reason for the war. I also had to secure a
group of one hundred and fifty SS prisoners, all cast from the same mould:
tough, vicious-looking, hard. I thought, Here's the reason we had to win the
war. God help us if we hadn't![4]

Benny Goodman echoes Grimes's views. He joined 102 Squadron after VE Day and was:

flying VIPs around Germany showing them the sites, the ruins, etcetera. I
still remember Cologne with the cathedral standing amidst the devastation
and shattered buildings – it really did bring it all home. This was the reality
of the bombing, but it had to be done. Hitler and his cohorts had to be
stopped. My conscience is clear. If you've ever visited a concentration camp, or
seen the gas chambers, seen the piles of bodies, you'd have no doubt at all
that we did the right thing.

'I despair when people say the bombing of the cities like Dresden or Cologne was wrong,' Tony Iveson said:

Somehow the minority has such a loud voice and is listened to and promoted
in the media, but the general public were with us all the way. That time was
unique in the history of war and in the history of aviation. Those four or five
years of bombing would never happen again – if it did, there would be no
civilisation – but that's how we waged war at the time; you were in a war to
the death. The Nazis wouldn't have treated us very kindly had they got
here. Talk to people in France or Denmark or Norway who lived through it
and hear what happened to them. It would have been a very different world
if we hadn't won the war.

Iveson was at the forefront of the campaign to have the longed-for memorial built as a lasting testament to the horrific sacrifice of his friends and colleagues. He died not long after it was unveiled.

However, for most of the veterans of 617 Squadron, the business of getting on with their lives, earning a living and raising a family took priority over justifying their wartime role. Colin Cole went back to his pre-war job. 'They had to give you your job back, so I returned to work as a clerk for local government. No one was interested in me at all. In fact, there was a sense of "Oh, he's back now, is he?" I was back from the war but there was no welcome. Indeed, they made me start back at the bottom again. No hero's welcome there!'[5]

Most of them rarely mentioned the war, even when they got together with old comrades. Johnny Johnson is one of only three men still alive who took part in the Dams raid. 'After the war, Joe McCarthy and I, with our wives, would meet up before each reunion, but we didn't really talk about the war,' he says. 'We discussed family life and where we were now, not where we had been. In the years after the war, even in the RAF, no one really mentioned the "Dambusters". It just wasn't of interest, or not until the film came out anyway.'[6]

'There was very little talk about the war,' agrees John Bell. 'I served in the RAF until 1977, and I can hardly remember the war being mentioned, though it is now, of course, as our numbers get fewer.'

Yet, while many veterans quietly got on with the business of rebuilding their lives, some could not forget. After the war, as his own private act of reparation for the damage he had helped to wreak on German cities and civilian lives, Nick Knilans devoted himself to constructive public service. 'I had made a vow that I would do some act to better mankind,' he said. 'I wanted to compensate for all the death and destruction I had helped create.' He worked as a teacher for twenty-five years, including two years as a Peace Corps volunteer in Nigeria, and also championed the improvement of the lives of young Mexican-Americans and became a counsellor in Californian prisons. Even after his retirement in 1978, he continued to support programmes for underprivileged youths right up until his death in 2012, aged ninety-four.[7]

Don Cheney never forgot the day he was shot down and three of his crew were killed. He was awarded the DFC for helping his crew to escape from their doomed aircraft, but seventy years later, even though he was completely blameless, he had never stopped reproaching himself for the three deaths. His daughter Jan recalls:

There was very little discussion about Dad's war experiences when I was growing up. Every August the fifth [the anniversary of his plane being shot down] Dad used to want to be alone and did not want to talk to anyone. I knew it had something to do with the war, but I only learned the true magnitude of what he had experienced in my late teens. Neither my mum or dad wanted to scare or disturb us by telling us earlier; and anyway, not talking about the war was very much a coping mechanism for Dad, who we know now has suffered from PTSD since coming back in 1944.

I never heard my mother talk about (and I have never asked her) what it was like to learn that my dad was missing in action and presumed dead. Not talking about it was also a way to cope for my mother, because it was when Dad talked about or was reminded of the war that he would experience severe anxiety and depression, and that was very hard on her and us kids. It would manifest by him not being able to get out of bed, being incredibly tense, stressed, unable to sleep, and unable to contribute or emotionally support my mother or us kids during those times. He was (and still is, at times) simply in survival mode. So, all in all, everyone thought the best way to deal with the war trauma was not to deal with it!

Dad said the worst thing he'd ever had to do in his life was to go and explain what had happened the day they were shot down to the mother of his wireless operator and friend, Reggie Pool, who was killed that day. That he had lost three of his friends and felt their deaths were his fault, occupied much of his thoughts throughout his life, however much he tried to suppress them. Dad told me not long ago that every night before he goes to sleep he thinks about the possibility that if he 'had only turned the plane a little more one way or another', they would not have been shot down and his friends would not have died that day.

My most vivid recollections of him when I was growing up were of him staring into space, lost in another world; I would have to shake his arm or hand, or sometimes take his face and turn it in the direction of my face, for him to come back to reality and pay attention to what I was trying to tell or show him. This happened all the time; the war robbed me of my dad, often at times when I really needed him to be there for me the most.[8]

Don died in August 2014 and was described as 'an incomparable role model, strikingly modest about himself but unabashedly boastful about the accomplishments of others'.[9] Before he died, speaking from his home in Canada, Cheney said, 'I'm so very proud of my role in the RCAF. I was flying alongside the best crew in the world and I still think of them every day. Those wartime experiences made me the man I am today.'[10]

★　　★　　★

As they grew older, some veterans made pilgrimages to the sites of 617 Squadron's most famous raids, laying wreaths on the graves of dead comrades or just reflecting on the events of all those years ago. Many, however, opted to stay away. A few years ago, Fred Sutherland was invited to visit the Eder dam he had helped to destroy, but he declined. 'I was really embarrassed at the thought of going there,' he says.

> *A lot of people were killed in the flooding. At the time we blew up the dam we never thought or worried about anything like that, but I think about it a bit differently now. We all lost a lot of friends back then, but it's only recently that I've found myself thinking about those days. I don't dwell on it, but I often think about the friends who died. It was such a loss of youth and promise and I often wonder what they'd have done, what they would have become. All that hope – just gone.*[11]

As one of the three surviving Dambusters – and the only Canadian – Sutherland was invited to Britain to commemorate the seventieth anniversary of the Dams raid in 2013, but again he politely declined the invitation. 'It's not much of a party,' he says, 'when there are so few of us left. My crew are all gone. Passed away. And most people don't really understand what we did anyway; they can't understand it, because they weren't there.'[12]

Perhaps it is the enduring scars left on the families of the dead that offer a lasting insight into the terrible cost. 'I received a letter from the son of Ken Gill, my navigator who was killed,' says Sydney Grimes. 'Meeting him brought it all very clearly home to me when he asked, "What was my father like?" I realised so many children like him never knew their fathers.'[13] Derek Gill is only one of tens of thousands of Bomber Command 'children' who would never know their fathers. He has spent many years researching his father's life. 'My mum never got over losing him and never remarried,' he says. 'I was so pleased to meet Sydney Grimes. He is the only person who could talk to me about my father's life in the RAF.'[14]

Despite the traumas of battle and the loss of friends that they all experienced, few of the men of 617 Squadron harboured any regrets about their role in the war. 'I look back now,' Frank Tilley says, 'and I am very proud of my service on 617 Squadron. I suppose my contribution to Bomber Command was quite small, a bit puny really, but I'm still proud of it. I was one of the lucky ones but I sometimes wonder why I survived and so many others didn't. I'm simply very grateful; I remember all those who didn't make it through.'

John Bell also remains very proud of his time on 617 Squadron:

It was an incredible period to be involved in the business of war, a time of huge change and development and achievement. I do think about the death and destruction of course, but I still think it was the only way to win the war. It would never happen again like that, but you can't use today's standards and reasoning to judge the all-out war of seventy years ago.

617 Squadron was also 'a huge part' of Johnny Johnson's life. 'I was lucky, I was privileged to have been with the right crew, in the right place, at the right time, to have been involved in the Dams raid,' he says, 'and I'm honoured that I was part of it all.'

But I don't like the thought of being a 'national treasure'! I said to my children a while ago, 'It's time we forgot about this war business,' but they said, 'You're part of history, Dad!' I don't want to be 'history'!

The brutality of my father when I was a child meant that I never imagined the life I could go on to have. We never, ever got on – he never mentioned my life in the RAF or my role in the Dams raid – but my childhood and my life in the RAF, alongside my life with Gwyn, made me the person I am. Gwyn was my rock when I was on ops, the steadying influence

that helped me focus on life outside the RAF. I've had the sort of life I never
dreamed about as a child, full of love, enjoyment, children, grandchildren and
great-grandchildren – a life of respect.[15]

Larry Curtis became a prominent member of the 617 Squadron
Association, and remained very close to Mick Martin, naming one
of his sons after him, and paying regular visits to the Sardinian
grave-site of Bob Hay, their bomb-aimer who was killed over the
Anthéor viaduct. Curtis said that the most important thing he
learned from the war was 'how good people were. A spirit was
built up that I'm sure would never have happened in any other
walk of life. The friendships we made in those days are still there.
And all I'd say is, if I had it all to do again, I'd do it again.'[16]

★   ★   ★

When the surviving veterans of 617 Squadron held a reunion in
1980, the former head of Bomber Command, Sir Arthur 'Bomber'
Harris, wrote a note to them. It began:

To all my old lags of 617 Squadron and any other bombers. Ignore any
sneers or smears of authors and those who find them the only means of selling
their wares, and buy instead the books written by Albert Speer and Doc.
Goebbels. From those two at the very centre of things from 38–45, but on the
wrong side, you will find irrefutable and ample first-class evidence that
the strategic bombing won the three main victories in the war:

    In the air. Because they forced the enemy on the defensive, building and
training practically nothing but fighters and fighter pilots in a despairing
attempt, which failed, to defend their Fatherland.

    On land. Because they gave the Allied armies absolute air supremacy
and blasted out of their way any and every attempt by the enemy to make a
successful stand or counter-attack.

    At sea. Because the bombers sank or destroyed twice as many enemy
capital ships as the Navy accounted for; sank, fatally damaged or destroyed
before they were launched, at least a dozen submarines for every one
the Navy scuppered; destroyed hundreds of small naval craft such as destroy-
ers, torpedo and gun boats, minesweepers, and trawlers etc. And finally

annihilated the enemy merchant marine on which they depended for vital ore supplies for industry.

The Germans had, Harris added:

900,000 fit soldiers on air defence and half the army's anti-tank guns, while Speer had 800,000 fit men trying and failing to keep the railways going, doubtless thousands more repairing urgent damage to war industry. If you know of any Allied army that took two million out of the enemy's fighting lines and half their vital anti-tank guns, I would be interested to hear about it. But that is all you old lags and loafers did for the pay you drew. No wonder authors and journalists find cause to smear you.

Harris signed off his letter to the 617 Squadron crews: 'Enjoy yourselves, and how well you deserve it!'[17]

Sources and Bibliography

(IWM: Imperial War Museum)

PERSONAL INTERVIEWS WITH THE AUTHOR
John Bell
Don, Gladys and Jan Cheney
Colin Cole
David Fellowes
Basil Fish
Lawrence 'Benny' Goodman
Sydney and Iris Grimes
Gerry Hobbs
Tony Iveson
George 'Johnny' Johnstone
Arthur Joplin
John Langston
Les Munro
Klaus Rohwedder
Fred Sutherland
Frank Tilley
Murray Valentine (Vagnolini)
Russell 'Rusty' Waughman

PERSONAL PAPERS AND DOCUMENTS
Air Secretary to L. Gumbley, 7 January 1949, 5/2/5451 PRS
Brookes, Mrs M., 'A WAAF's Tale', 2 July 1993, IWM 93/22/1
Buhner, Werner, Civilian Accounts of the Bombing of the Arnsberg
    Viaduct

'Damn Busters' dinner menu, RAF Museum, Hendon, AC 96/12

Frick, Axel, Bielefeld Viaduct Civilian Accounts

Gill, Ken, letters and correspondence with relatives

Gumbley, Bernard 'Barney', letters and correspondence with relatives

Gumbley, John, letter to Sydney Grimes, 2005

Harris, Sir Arthur 'Bomber', letter to 617 Squadron Reunion, 18 April 1980, RAF Museum, Hendon

Hebbard, Loftus, correspondence with relatives

Hobbs, G. H., IWM Documents 1839

Holt, A. A., IWM Documents 8597

Knilans, Nick, 'A Yank in the RCAF', private memoir, RAF Museum

Muirhead, Campbell, diary

Pryor, John, personal recollections supplied by Robert Owen, Official 617 Squadron Association Historian

Wakefield, H. E., DFC, IWM Documents 15411

Walsh, J., IWM Documents 12812

Wilshire, Cyril 'Charlie', correspondence with relatives

AUDIO RECORDINGS

Bickley, Wilfred George, IWM Sound 14588

Briars, Ralph Algernon, IWM Sound 13924

Curtis, Lawrence Wesley, IWM Sound 92111

Cheshire, Geoffrey Leonard, IWM Sound 9861

Drobinski, Boleslaw Henryk, IWM Sound 12892

Hamilton, Malcolm Lennox, IWM Sound 18264

Hobday, Harold Sydney, IWM Sound 7298

Johnson, Edward Cuthbert, IWM Sound 8204

Kearns, Richard Stansfield Derek, IWM Sound 9302

Knights, Robert Edgar, IWM Sound 9208

Munro, John Leslie, IWM Sound 33077

Poore, Arthur Frank, IWM Sound 20261

Reid, William 'Bill', IWM Sound 4993

Sanders, John Aelred, IWM Sound 14803

Shannon, David John, IWM Sound 8177

Tait, James Brian 'Willie', IWM Sound 2519

Ware, David R., IWM Sound 24932
Watts, Frederick Henry Arthur, IWM 21029

PERIODICALS

Après Moi! is the 617 Squadron Association newsletter
Memorial Flight is the journal of Lincolnshire's Lancaster Association

Allett, Tom, 'The Bright Sparks', *Memorial Flight*, Spring 2007
Allett, Tom, 'First Op *Tirpitz*', *Memorial Flight*, Spring 2009
'Bill & Ben, Mosquito Men', *Memorial Flight*, 2009
Bennett, Tom, 'Operation Taxable', *Flypast*, No. 40, November 1984
Burgess, Colin, '617 Squadron: The Later War Operations', *Transit Magazine*, Australia 1986
Chorlton, Martin, 'A Home for Heroes', *Memorial Flight*, Spring 2009
Cooper, Dennis, 'A Visit to Wartime Russia or *Tirpitz* Ahoy!', *Après Moi!*, Winter 2009/10
Cotterell, Anthony, 'Did I Ever Tell You about My Operation?', *War: Frankfurt Revisited*, Army Bureau of Current Affairs, No. 62, 22 January 1944, London
Daily Telegraph, 19 May 1943
'Dam-buster Gibson to Get VC', *Daily Mirror*, 27 May 1943
'Dropped 10-Tonner on Huns', *Illustrated London News*, 24 March 1945
Eeles, Ron, 'My Recollections of a Night Bombing Raid upon Mailly le Camp, France', www.49squadron.co.uk
Everyone's War: The Journal of the Second World War Experience, No. 22, Autumn/Winter 2010
Gellhorn, Martha, 'The Bomber Boys', *Collier's Weekly*, 17 June 1944
Goodman, Benny, 'RAF Against the E-Boats: Rotterdam 29 December 1944', *Après Moi!*, Winter 2009/10
Goulding, Ossian, 'Heart of Berlin is Paralysed', *Daily Telegraph*, 25 November 1943
Hobbs, Gerry, 'Glad to be Back', *Après Moi!*, Spring 2008
Hobday, Sidney, 'Two Months in Occupied Europe', *Lloyds' Log*, October 1946
'A Home for Heroes: The History of Woodhall Spa', *Memorial Flight*, 2009

'Major Nick Knilans', obituary, *Daily Telegraph*, 27 June 2012

Margry, Karel, 'The Bielefeld Viaduct', *After the Battle*, No. 79, 1993

Morris, Richard, 'The Legacy of the Dambusters', *BBC History Magazine*, May 2013

Morris, Richard, 'Prosopography – A Special Group of Men', *Après Moi!*, Summer 2011

Munro, Les, 'A Rude Awakening', *Après Moi!*, Spring 2007

Noble, Vernon, 'They Call This Man "Killer"', *Daily Mirror*, 22 May 1943

Poore, Arthur, 'Personal Recollections', *Après Moi!*, Summer 2010

Shortland, Jim, '*Tirpitz* Survivors Recall Raids', *Après Moi!*, Autumn 2006

'Sinking of *Tirpitz*', New Zealand Press Association, 28 November 1944

'Specsavers 62 Years On!', *Après Moi!*, Autumn 2007

'Squadron Leader Tony Iveson', obituary, *Daily Telegraph*, 10 November 2013

Stopes-Roe, Mary, 'Memories', *Après Moi!*, Summer 2013

Storey, Ron, 'Mailly le Camp, "Like Falling off a Log"', *People's War*, BBC, Article ID A4292804

'"Tallboy" Wallis', *Daily Express*, 15 November 1944

Trevor, Hugh, 'Blockhouse Buster', *Flypast*, October 2010

Ward, Arthur, 'With Tait to the *Tirpitz*', *Après Moi!*, Summer 2010

'Warrant Officer Tom McLean', *Daily Telegraph*, 15 August 2011

Welden, Yvonne, 'Mailly le Camp – Sidney Lipman', *People's War*, BBC, Article ID A6660371

Whitehead, Chris, 'Knights of the Air', *Aviation News*, November 2003

Zuba, Alfred, '*Tirpitz*: A German Sailor's Eyewitness Report', *Après Moi!*, Winter 2009/10

BOOKS

Barbier, Mary, *D-Day Deception: Operation Fortitude and the Normandy Invasion*, Greenwood Publishing Group, Westport, 2007

Bennett, Tom, *617 Squadron: The Dambusters at War*, Patrick Stephens, Wellingborough, 1987

Bishop, Patrick, *Target Tirpitz*, Harper Press, London, 2012

Blundell, Harold M. 'Nobby', *They Flew from Waddington! 463-467*

Lancaster Squadrons, Royal Australian Air Force: A Brief History, 463–467 Squadrons Association, NSW Tour Committee, Sydney, 1975

—, *463-467 Squadrons R.A.A.F.*, 463-467 Squadrons Association, 1995

Boyle, Andrew, *No Passing Glory: The Full & Authentic Biography of Group Captain Cheshire*, Collins, London, 1955

Chandler, Chan, *Tail Gunner: 98 Raids in World War II*, Airlife Publishing, Bury St Edmunds, 2001

Cooper, Alan, *Beyond the Dams to the* Tirpitz, William Kimber, London, 1983

—, *The Dambusters Squadron: Fifty Years of 617 Squadron RAF*, Arms & Armour Press, London, 1993

Currie, Jack, *Battle under the Moon*, AirData Publications, Wilmslow, 1995

Delve, Ken, and Peter Jacobs, *The Six-Year Offensive*, Arms & Armour 1992

Eaton, E. A., *Two Friends: Two Different Hells*, ReCall Publications, Northallerton, 2002

Foster, Charles, *Breaking the Dams: The Story of Dambuster David Maltby and His Crew*, Pen & Sword, Barnsley, 2008

Grayling, A. C., *Among the Dead Cities*, Bloomsbury, London, 2006

Hastings, Max, *Bomber Command*, Pan, London, 1999

Heathcote, Blake, *Testaments of Honour*, Doubleday, London, 2002

Hesketh, Roger, *Fortitude*, Overlook Press, New York, 2000

Higgs, Colin, and Bruce Vigar, *Voices in Flight*, Pen & Sword, Barnsley, 2013

Hobday, Sydney, *Two Months in Occupied Europe* (publisher/date unknown)

Humphries, Harry, *Living with Heroes: The Dambusters*, The Erskine Press, Eccles, Norwich, 2003

Iveson, Tony, *Lancaster: The Biography*, André Deutsch, London, 2009

Kellow, Bob, *Paths to Freedom*, Kellow Corporation, Winnipeg, 1992

Lambert, Max, *Night after Night*, HarperCollins, New Zealand, 2007

Le Maner, Yves, *Mimoyecques: Site of 'London Cannon' (1943–1945)*, La Coupole Editions, Nord-Pas-de-Calais, 2011

Levine, Joseph, *Operation Fortitude*, Collins, London, 2012

Middlebrook, Martin, and Chris Everitt, *The Bomber Command War Diaries*, Midland Publishing, Birmingham, 1985

Morris, Richard, *Cheshire: The Biography of Leonard Cheshire*, Penguin, Harmondsworth, 2001

—, *Guy Gibson*, Viking, London, 1994

Murray, Iain R., *Dam Busters: Owners' Workshop Manual*, Haynes Publishing, Yeovil, 2011

Nichol, John, *The Red Line*, William Collins, London, 2013

Nichol, John and Tony Rennell, *Tail-End Charlies*, Penguin, Harmondsworth, 2005

Pasmore, Anthony, and Norman Parker, *Ashley Walk: Its Bombing Range, Landscape and History*, New Forest Research & Publication Trust, Lyndhurst, 2006

Probert, Air Commodore Henry, *Bomber Harris*, Greenhill Books, Barnsley, 1998

Rowley, Clive, *Dambusters: The Most Daring Raid in the RAF's History*, Mortons Media Group, Horncastle, 2013

Russell of Liverpool, Lord, *Scourge of the Swastika: A Short History of Nazi War Crimes*, Greenhill Books, Barnsley, 2005

Simpson, Tom, *Lower than Low*, Libra, Victoria, 1995

Ward, Chris, *Dambusters: The Forging of a Legend*, Pen & Sword, Barnsley, 2009

Zaloga, Steven J., *German V-Weapon Sites 1943–45*, Osprey, Oxford, 2007

TELEVISION PROGRAMMES

The Dambusters' Great Escape, Channel 4, broadcast Sunday 30 March 2014

What the Dambusters Did Next, Channel 5, broadcast Monday 26 May 2014

WEBSITES

Kevin James Biltoff ID Number S01771 http://static.awm.gov.au/audio/S01771--1-.mp3

Marc Buggeln (transl. Stephen Pallavicini), *The United States Holocaust Memorial Museum Encyclopedia of Camps and Ghettos, 1933–1945*

(Center for Advanced Holocaust Studies, United States Holocaust Memorial Museum) http://www.ushmm.org/wlc/en/article.php? ModuleId=10007390

http://histru.bournemouth.ac.uk/Oral_History/Talking_About_ Technology/radar_research/fooling_the_enemy.html

http://www.flyingbombsandrockets.com/storeys_summary.html

Charles Foster, Dambusters Blog, 30 January 2014 http://dam bustersblog.com/category/fred-sutherland/

http://www.bombercommandmuseum.ca/commandlosses.html

Notes

CHAPTER I THE DAMS

1　Fred Sutherland, interview with the author
2　George 'Johnny' Johnson, interview with the author
3　ibid.
4　Clive Rowley, *Dambusters*
5　ibid.
6　Quoted in Charles Foster, *Breaking the Dams*
7　Fred Sutherland, interview with the author
8　Chris Ward, *Dambusters: The Forging of a Legend*
9　ibid.
10　ibid.
11　Harry Humphries, *Living with Heroes*
12　Mrs M. Brookes, 'A WAAF's Tale'
13　George 'Johnny' Johnson, interview with the author
14　ibid.
15　Richard Morris, 'Prosopography – A Special Group of Men', *Après Moi!*, Summer 2011
16　Lawrence Wesley Curtis, IWM Sound 92111
17　Clive Rowley, *Dambusters*
18　George 'Johnny' Johnson, interview with the author
19　Quoted in Charles Foster, *Breaking the Dams*
20　*Daily Telegraph*, 19 May 1943
21　Clive Rowley, *Dambusters*
22　ibid.
23　Quoted in Richard Morris, 'The Legacy of the Dambusters'
24　Mary Stopes-Roe, interview with the author
25　Tom Simpson, *Lower than Low*

26 Harry Humphries, *Living with Heroes*
27 George 'Johnny' Johnson, interview with the author
28 Tom Simpson, *Lower than Low*
29 George 'Johnny' Johnson, interview with the author
30 ibid.
31 Tom Simpson, *Lower than Low*
32 ibid.
33 'Damn Busters' dinner menu, RAF Museum, Hendon
34 Martin Middlebrook and Chris Everitt, *The Bomber Command War Diaries*
35 Charles Foster, *Breaking the Dams*
36 E. A. Eaton, *Two Friends: Two Different Hells*
37 Tom Simpson, *Lower than Low*
38 Alan Cooper, *Beyond the Dams to the* Tirpitz
39 Lawrence Wesley Curtis, IWM Sound 92111
40 'Squadron Leader Larry Curtis', obituary, *Daily Telegraph*, 29 June 2008
41 George 'Johnny' Johnson, interview with the author
42 ibid.
43 David Fellowes, interview with the author
44 Frank Tilley, interview with the author
45 E. A. Eaton, *Two Friends: Two Different Hells*
46 Quoted in Tom Simpson, *Lower than Low*

CHAPTER 2 WHAT NEXT?

 1 George 'Johnny' Johnson, interview with the author
 2 National Archives H385, CAS 2944
 3 Tom Simpson, *Lower than Low*
 4 Harry Humphries, *Living with Heroes*
 5 As with many theories advanced in the post-war years, this is disputed by some
 6 Lawrence Wesley Curtis, IWM Sound 92111
 7 George 'Johnny' Johnson, interview with the author
 8 Lawrence Wesley Curtis, IWM Sound 92111
 9 Tom Simpson, *Lower than Low*
10 Chris Ward, *Dambusters: The Forging of a Legend*

11 Sidney Hobday, *Two Months in Occupied Europe*. Account provided by Charles Foster
12 Bob Kellow, *Paths to Freedom*
13 Fred Sutherland, interview with the author
14 E. A. Eaton, *Two Friends: Two Different Hells*
15 Sidney Hobday, *Two Months in Occupied Europe*
16 George 'Johnny' Johnson, interview with the author
17 ibid.
18 Lawrence Wesley Curtis, IWM Sound 92111
19 Peter Spoden, interview for *What the Dambusters Did Next*, Channel 5
20 David Fellowes, interview with the author
21 Lawrence Wesley Curtis, IWM Sound 92111
22 Sidney Hobday, *Two Months in Occupied Europe*
23 ibid.
24 This edited version of the letter appears in Charles Foster, *Breaking the Dams*
25 ibid., edited version

Chapter 3 Press On, Regardless

1 Quoted in A. C. Grayling, *Among the Dead Cities*
2 Chan Chandler, *Tail Gunner*
3 Lawrence Wesley Curtis, IWM Sound 92111
4 Malcolm Lennox Hamilton, IWM Sound 8264
5 George 'Johnny' Johnson, interview with the author
6 Tom Simpson, *Lower than Low*. There is some disagreement about whether Wallis was actually on the base at this time, but this is Simpson's recollection
7 '"Tallboy" Wallis', *Daily Express*, 15 November 1944
8 Lawrence Wesley Curtis, IWM Sound 92111
9 Sydney Grimes, interview with the author
10 Wilfred George Bickley, IWM Sound 14588
11 Sydney Grimes, interview with the author
12 Nick Knilans, 'A Yank in the RCAF'
13 Tom Simpson, *Lower than Low*
14 Nick Knilans, 'A Yank in the RCAF'

15 Loftus Hebbard, details supplied by his son, Bruce Hebbard
16 Martha Gellhorn, 'The Bomber Boys', *Collier's Weekly*,
    17 June 1944
17 'A Home for Heroes', *Memorial Flight*, 2009
18 Chan Chandler, *Tail Gunner*
19 ibid.
20 Lawrence Wesley Curtis, IWM Sound 92111
21 Bruce 'Buck' Buckham, DSO, DFC, Australian War Memorial,
    IWM ID Number S01670

CHAPTER 4 DEATH OR GLORY

1 Tom Bennett, personal recollections supplied by Rob Owen,
   Official 617 Squadron Association Historian
2 Wilfred George Bickley, IWM Sound 14588
3 'Squadron Leader Larry Curtis', obituary, *Daily Telegraph*,
   29 June 2008
4 Lawrence Wesley Curtis, IWM Sound 92111
5 David Fellowes, interview with the author
6 Anthony Cotterell, 'Did I Ever Tell You about My
   Operation?'
7 Nick Knilans, 'A Yank in the RCAF'
8 George 'Johnny' Johnson, interview with the author
9 ibid.
10 Malcolm Lennox Hamilton, IWM Sound 8264
11 Ossian Goulding, 'Heart of Berlin Is Paralysed'
12 J. Walsh, IWM Documents 12812
13 John Bell, interview with the author
14 Lawrence Wesley Curtis, IWM Sound 92111
15 Les Munro, interview with the author
16 George 'Johnny' Johnson, interview with the author
17 Harry Humphries, *Living with Heroes*
18 ibid.
19 Don Cheney, interview with the author. Sadly, Don died a
    few months after giving the interview
20 For a fuller account of Tom McLean's extraordinary exploits,
    see Tom Bennett, *617 Squadron*; 'Warrant Officer

Tom McLean', obituary, *Daily Telegraph*, 15 August 2011; www.worldwarbirdnews.com; *Après Moi!*, Summer 2011

21 *Daily Mirror*, 22 May 1943
22 Tom Bennett, *617 Squadron*
23 Wilfred George Bickley, IWM Sound 14588
24 Nick Knilans, 'A Yank in the RCAF'
25 'Major Nick Knilans', obituary, *Daily Telegraph*, 27 June 2012

CHAPTER 5 SPRING 1944
1 George 'Johnny' Johnson, interview with the author
2 John Pryor, Juvisy, 18 April 1944, article supplied by Rob Owen, Official 617 Squadron Association Historian
3 Tom Bennett, article supplied by Rob Owen, Official 617 Squadron Association Historian
4 Chris Whitehead, *Knights of the Air*
5 Wilfred George Bickley, IWM Sound 14588
6 Tom Bennett, article supplied by Rob Owen, Official 617 Squadron Association Historian
7 Richard Morris, *Cheshire*
8 Tom Bennett, article supplied by Rob Owen, Official 617 Squadron Association Historian

CHAPTER 6 THE END OF THE BEGINNING
1 Some records show differing dates
2 Tom Bennett, 'Operation Taxable', *Flypast*, No. 40, November 1984
3 Ron Storey, 'Mailly le Camp, "Like Falling off a Log"', *People's War*, BBC
4 Rusty Waughman, personal account and interview with the author
5 Ron Storey, 'Mailly le Camp, "Like Falling off a Log"', *People's War*, BBC
6 Ron Eeles, 'My Recollections'
7 Ron Storey, 'Mailly le Camp, "Like Falling off a Log"', *People's War*, BBC

8 'Bill & Ben, Mosquito Men', *Memorial Flight*, Spring 2009

9 Ron Eeles, 'My Recollections'

10 Quoted in Yvonne Weldon, 'Mailly le Camp – Sidney Lipman', *People's War*, BBC

11 Rusty Waughman account

12 Alan Cooper, *Beyond the Dams to the* Tirpitz

13 Ron Storey, 'Mailly le Camp, "Like Falling off a Log"', *People's War*, BBC

14 Tom Bennett, 'Operation Taxable', *Flypast*, No. 40, November 1984

15 Arthur Poore, 'Personal Recollections', *Après Moi!,* Summer 2010

16 Don Cheney, interview with the author

17 Alan Jordan, quoted in Colin Burgess, '617 Squadron: The Later War Operations'

18 Tom Bennett, 'Operation Taxable', *Flypast*, No. 40, November 1984

19 Gerry Hobbs, interview with the author

20 Nick Knilans, 'A Yank in the RCAF'

21 Tom Bennett, 'Operation Taxable', *Flypast*, No. 40, November 1984

CHAPTER 7 THE FIGHT GOES ON

1 Martin Middlebrook and Chris Everitt, *The Bomber Command War Diaries*

2 Malcolm Lennox Hamilton, IWM Sound 8264

3 ibid.

4 ibid.

5 ibid.

6 'Specsavers 62 Years On!', *Après Moi!*, Autumn 2007

7 Malcolm Lennox Hamilton, IWM Sound 8264

8 'Specsavers 62 Years On!', *Après Moi!*, Autumn 2007

9 ibid.

10 Malcolm Lennox Hamilton, IWM Sound 8264

CHAPTER 8 TERROR WEAPONS

1 George 'Johnny' Johnson, interview with the author

2 Nick Knilans, 'A Yank in the RCAF'

3 http://archive.iwm.org.uk/upload/package/4/dday/pdfs/ VWeaponsCampaign.pdf

4 John Pryor, Mimoyecques V-3, 6 July 1944, personal account supplied by Rob Owen, Official 617 Squadron Association Historian

5 Mimoyecques statistics from Yves le Maner, Mimoyecques

6 Quoted in Tony Iveson, *Lancaster: The Biography*

7 Tony Iveson, interview with the author

8 http://www.flyingbombsandrockets.com/storeys_summary.html

9 ibid.

10 G. H. Hobbs, personal account and interview with the author

11 John Pryor, Wizernes, 17 and 20 July 1944, personal account supplied by Rob Owen, Official 617 Squadron Association Historian

12 G. H. Hobbs, personal account and interview with the author

13 Lawrence Wesley Curtis, IWM Sound 92111

14 Arthur Ward, 'With Tait to the *Tirpitz*', *Après Moi!*, Summer 2010

15 Frank Tilley, interview with the author

16 Nick Knilans, 'A Yank in the RCAF'

17 Malcolm Lennox Hamilton, IWM Sound 8264

18 From V-2 factory fact sheets – Wizernes: La Coupole

19 This story is told in different ways by those who were there – it is likely that Tait was simply indicating the point to target rather than the notion of actually bombing his own aircraft

CHAPTER 9 MAC'S GONE!

1 Don Cheney, personal account and interview with the author

2 Tony Iveson, *Lancaster: The Biography*

3 *London Gazette*, Supplement 36285, 10 December 1943, https://www.thegazette.co.uk/London/issue/36285/ supplement/5435 accessed 12 May 2014

4 Bill Reid, from a document titled 'Irvin Parachute Editorial'

5 Bill Reid, handwritten account provided by Geoff Robertson

6 ibid.

7 Murray Vagnolini, interview with the author

CHAPTER 10 LIFE AND DEATH ON 617 SQUADRON

1 Anthony Cotterell, 'Did I Ever Tell You about My Operation?'
2 Les Munro, 'A Rude Awakening', *Après Moi!*, Spring 2007
3 Tom Bennett, *617 Squadron*
4 John Pryor, Brest, 12–13–14 August 1944, personal account supplied by Rob Owen, Official 617 Squadron Association Historian
5 Arthur Ward, 'With Tait to the *Tirpitz*', *Après Moi!*, Summer 2010
6 John Pryor, Brest, 12–13–14 August 1944, personal account supplied by Rob Owen, Official 617 Squadron Association Historian

CHAPTER 11 THE BEAST

1 Some witnesses maintain Wallis was not even on base that day – such is the confusion of wartime recollections.
2 Bruce 'Buck' Buckham, Australian War Memorial, ID Number S01670, and research by Nobby Blundell, edited by Bruce Buckham, http://forums.diecast-aviation.eu/showthread. php?t=20887; see also Harold M. 'Nobby' Blundell, *463-467 Squadrons R.A.A.F.* and *They Flew from Waddington! 463-467 Lancaster Squadrons, Royal Australian Air Force*
3 Malcolm Lennox Hamilton, IWM Sound 8264
4 Frank Tilley, interview with the author
5 http://www.smh.com.au/comment/obituaries/bomber-pilot-captured-tirpitzs-demise-20110816-1iw9r.html#ixzz2vwFjnXHF
6 George 'Johnny' Johnson, interview with the author
7 Bruce 'Buck' Buckham, DSO, DFC, Australian War Memorial, IWM ID Number S01670
8 Frank Tilley, interview with the author
9 Alan Cooper, *Beyond the Dams*
10 Basil Fish, interview with the author
11 Malcolm Lennox Hamilton, IWM Sound 8264
12 Sydney Grimes, interview with the author
13 Malcolm Lennox Hamilton, IWM Sound 8264
14 Harry Humphries, *Living with Heroes*

15 Tom Bennett, *617 Squadron*

16 Arthur Ward, 'With Tait to the *Tirpitz*', *Après Moi!*, Summer 2010

17 Malcolm Lennox Hamilton, IWM Sound 8264

18 Alf Jackson, quoted in Colin Burgess, '617 Squadron: The Later War Operations'

19 Lawrence Wesley Curtis, IWM Sound 92111

20 ibid.

21 Malcolm Lennox Hamilton, IWM Sound 8264

22 ibid.

23 Alan Cooper, *Beyond the Dams*

24 Bruce 'Buck' Buckham, DSO, DFC, Australian War Memorial, ID Number S01670

25 Arthur Ward, 'With Tait to the *Tirpitz*', *Après Moi!*, Summer 2010

26 Dennis Cooper, 'A Visit to Wartime Russia or *Tirpitz* Ahoy!', *Après Moi!*, Winter 2009/10

27 Bruce 'Buck' Buckham, DSO, DFC, Australian War Memorial, ID Number S01670

28 Arthur Ward, 'With Tait to the *Tirpitz*', *Après Moi!*, Summer 2010

29 Dennis Cooper, 'A Visit to Wartime Russia or *Tirpitz* Ahoy!', *Après Moi!*, Winter 2009/10

30 Chris Whitehead, *Knights of the Air*

31 Klaus Rohwedder, interview for *What the Dambusters Did Next*, Channel 5

32 Wilfred George Bickley, IWM Sound 14588

33 ibid.

34 Bruce 'Buck' Buckham, DSO, DFC, Australian War Memorial, ID Number S01670

35 Lawrence Wesley Curtis, IWM Sound 92111

36 Arthur Ward, 'With Tait to the *Tirpitz*', *Après Moi!*, Summer 2010

37 Alan Cooper, *Beyond the Dams*

38 Lawrence Wesley Curtis, IWM Sound 92111

39 Klaus Rohwedder, interview with the author

40 Chris Ward, *Dambusters*
41 Klaus Rohwedder, interview with the author

CHAPTER 12 'WHAT HAVE YOU BEEN DOING TODAY?'

1 Quoted in Richard Morris, *Guy Gibson*
2 Frank Tilley, interview with the author
3 Basil Fish, interview with the author
4 Loftus Hebbard, information provided by Bruce Hebbard
5 Alan Cooper, *From the Dams to the* Tirpitz
6 Martin Middlebrook and Chris Everitt, *The Bomber Command War Diaries*
7 Correspondence with John Saunders, great-nephew of Bruce Hosie, and Max Lambert, *Night after Night*
8 http://www.traffordwardead.co.uk/index.php?sold_id=s%3A17%3A%22594%2Cstretford_ww2%22%3B&letter=W&place=&war=&soldier=Wyness+DFC
9 Tony Iveson, interview with the author
10 Quoted in Colin Burgess, '617 Squadron: The Later War Operations'
11 Colin Burgess, '617 Squadron: The Later War Operations'
12 Tony Iveson, *Lancaster: The Biography*
13 Dennis Cooper, 'A Visit to Wartime Russia or *Tirpitz* Ahoy!', *Après Moi!*, Winter 2009
14 Tommy Trebilcock, in Colin Burgess, '617 Squadron: The Later War Operations'
15 Arthur Ward, 'With Tait to the *Tirpitz*', *Après Moi!*, Summer 2010
16 Letter provided by John Saunders
17 Lord Russell of Liverpool, *Scourge of the Swastika*
18 Max Lambert, *Night after Night*
19 John Saunders, letter to the author, April 2014
20 Willie Tait, in Colin Burgess, '617 Squadron: The Later War Operations'

CHAPTER 13 BACK TO THE *TIRPITZ*

1　Benny Goodman, interview with the author
2　Frank Tilley, interview with the author
3　ibid.
4　Bob Barry, quoted in Colin Burgess, '617 Squadron: The Later War Operations'
5　Arthur Ward, 'With Tait to the *Tirpitz*', *Après Moi!*, Summer 2010
6　Sydney Grimes, interview with the author
7　Malcolm Hamilton, IWM Sound 8264
8　Iris Grimes, interview with the author
9　Bruce 'Buck' Buckham, DSO, DFC, Australian War Memorial, IWM ID Number S01670
10　George 'Johnny' Johnson, interview with the author
11　Campbell Muirhead, diary
12　Bernard 'Barney' Gumbley, letters supplied by his nephew, John Gumbley
13　Frank Tilley, interview with the author
14　Porjus (Sweden) town archive
15　Colin Burgess, '617 Squadron: The Later War Operations'
16　Tom Bennett, *617 Squadron*
17　Colin Burgess, '617 Squadron: The Later War Operations'
18　ibid.
19　Harry Humphries, *Living with Heroes*
20　Alan Cooper, *Beyond the Dams to the* Tirpitz
21　Chris Ward, *Dambusters*
22　Malcolm Lennox Hamilton, IWM Sound 8264
23　Bruce 'Buck' Buckham, DSO, DFC, Australian War Memorial, IWM ID Number S01670
24　Tom Bennett, *617 Squadron*
25　Colin Cole, interview with the author
26　Murray Vagnolini, interview with the author
27　Tom Allett, 'First Op *Tirpitz*'
28　Kurt Schultze, interview for *The Dambusters' Great Escape*, Channel 4
29　Lawrence Wesley Curtis, IWM Sound 92111

30 Phil Tetlow, interview for *The Dambusters' Great Escape*, Channel 4

31 Colin Cole, quoted by Tom Allett in 'First Op *Tirpitz*'

32 Frank Tilley, interview with the author

33 ibid.

34 Patrick Bishop, interview for *The Dambusters' Great Escape*, Channel 4

35 Bruce 'Buck' Buckham, DSO, DFC, Australian War Memorial, IWM ID Number So1670

36 Colin Cole, interview with the author

37 Alfred Zuba, '*Tirpitz*: A German Sailor's Eyewitness Report', *Après Moi!*, Winter 2009/10

38 Quoted in Jim Shortland, '*Tirpitz* Survivors Recall Raids', *Après Moi!*, Autumn 2006

39 Frank Tilley, interview with the author

40 Tony Iveson, interview with the author

41 Frank Tilley, interview with the author

42 Frank Tilley, interview with the author

43 Colin Cole, interview with the author.

44 Bruce 'Buck' Buckham, DSO, DFC, Australian War Memorial, IWM ID Number So1670

45 ibid.

46 Frank Tilley, interview with the author

47 Chris Ward, *Dambusters*

48 Chan Chandler, *Tail Gunner*

49 Sydney Grimes, interview with the author

50 Alfred Zuba, '*Tirpitz*: A German Sailor's Eyewitness Report', *Après Moi!*, Winter 2009/10

51 Quoted in Alan Cooper, *Beyond the Dams*

52 Quoted in Alan Cooper, *Beyond the Dams*

53 Jim Shortland, '*Tirpitz* Survivors Recall Raids', *Après Moi!*, Autumn 2006

54 Alfred Zuba, '*Tirpitz*: A German Sailor's Eyewitness Report', *Après Moi!*, Winter 2009/10

55 Kurt Schultze, interview for *The Dambusters' Great Escape*, Channel 4

56 Alan Cooper, *Beyond the Dams*

57 Kurt Schultze, interview for *The Dambusters' Great Escape*,
    Channel 4

58 Bob Barry, quoted in Colin Burgess, '617 Squadron: The Later
    War Operations'

59 Arthur Ward, 'With Tait to the *Tirpitz*', *Après Moi!*, Summer 2010

60 ibid.

61 Reproduced from research by Nobby Blundell.
    http://forums.diecast-aviation.eu/showthread.php?t=20887;
    Harold M. 'Nobby' Blundell, *467-463 Squadrons R.A.A.F.* and
    *They Flew from Waddington! 463-467 Lancaster Squadrons, Royal
    Australian Air Force*

62 http://www.smh.com.au/comment/obituaries/bomber-pilot-
    captured-tirpitzs-demise-20110816-1iw9r.html#ixzz2vwI8GmM9

63 Arthur Ward, 'With Tait to the *Tirpitz*', *Après Moi!*, Summer
    2010

64 Frank Tilley, interview with the author

65 Private papers of J. B. Tait, quoted in Patrick Bishop, *Target
    Tirpitz*

66 ibid.

67 Bob Barry, quoted in Colin Burgess, '617 Squadron: The Later
    War Operations'

68 Bernard 'Barney' Gumbley, letters supplied by John Gumbley

CHAPTER 14 THE LAST CHRISTMAS?

 1 Arthur Joplin, interview with the author

 2 Frank Tilley, interview with the author

 3 Sydney Grimes, interview with the author

 4 Benny Goodman, interview with the author

 5 Ken Gill, personal letters supplied by his son Derek Gill

 6 Tony Iveson, *Lancaster: The Biography*

 7 ibid., and interview with the author

 8 Murray Vagnolini, interview with the author

 9 John Pryor, Bergen, 12 Jan 1944, personal account supplied by
    Rob Owen, Official 617 Squadron Association Historian

10 Colin Cole, interview with the author

11 Quoted in Ken Delve and Peter Jacobs, *The Six-Year Offensive*

12 ibid.

13 Karel Margry, 'The Bielefeld Viaduct', *After the Battle*, No. 79, 1993

14 ibid.

15 John Langston, interview with the author

16 Colin Cole, interview with the author

17 John Langston, interview with the author

18 Sydney Grimes, interview with the author

19 Murray Vagnolini, interview with the author

20 'Dropped 10-Tonner on Huns', *Illustrated London News*, 24 March 1945

21 Murray Vagnolini, interview with the author

22 H. E. Wakefield DFC, IWM Documents 15411

23 Karel Margry, 'The Bielefeld Viaduct', *After the Battle*, No. 79, 1993

24 Axel Frick, Bielefeld Viaduct Civilian Accounts

25 H. E. Wakefield DFC, IWM Documents 15411

26 Don Day, quoted in Colin Burgess, '617 Squadron: The Later War Operations'

27 Benny Goodman, interview with the author

28 Sydney Grimes, interview with the author

29 Colin Burgess, '617 Squadron: The Later War Operations'

30 Werner Buhner, Civilian Accounts of the Bombing of the Arnsberg Viaduct

31 ibid.

32 Bob Barry, quoted in Colin Burgess, '617 Squadron: The Later War Operations'

33 Letter, Air Secretary to L. Gumbley, 7 January 1949

34 John Gumbley, letter to Sydney Grimes, 2005

35 Sydney Grimes, interview with the author

Chapter 15 The Final Days

1 http://www.ushmm.org/wlc/en/article.php?ModuleId=10007390

2 Email correspondence with Michelle Callan

3 Bob Barry, quoted in Colin Burgess, '617 Squadron: The Later War Operations'

4 ibid.

5 Email correspondence with Michelle Callan

6 Murray Vagnolini, interview with the author

7 David R. Ware, IWM Sound 24932

8 ibid.

9 Boleslaw Henryk Drobinski, IWM Sound 12892

10 Tony Iveson, interview with the author

11 Martin Middlebrook and Chris Everitt, *The Bomber Command War Diaries*

12 Les Munro, interview with the author

13 Chris Ward, *Dambusters: The Forging of a Legend*

14 http://www.bombercommandmuseum.ca/commandlosses.html

Chapter 16 Counting the Cost

1 Frank Tilley, interview with the author

2 Hansard, 6 March 1945

3 Henry Probert, *Bomber Harris*

4 Sydney Grimes, interview with the author

5 Colin Cole, interview with the author

6 George 'Johnny' Johnson, interview with the author

7 'Major Nick Knilans', obituary, *Daily Telegraph*, 27 June 2012

8 Correspondence with Jan Cheney

9 http://www.legacy.com/obituaries/ottawacitizen/obituary.aspx?n=donald-cheney&pid=172298644

10 Don Cheney, interview with the author

11 Fred Sutherland, interview with the author

12 http://news.nationalpost.com/2013/05/17/a-real-suicide-run-70-years-after-the-dambusters-raid-an-airman-tells-remarkable-tale/

13 Sydney Grimes, interview with the author

14 Derek Gill, correspondence with the author

15 George 'Johnny' Johnson, interview with the author

16 Lawrence Wesley Curtis, IWM Sound 92111

17 Letter from Sir Arthur 'Bomber' Harris, 18 April 1980, RAF Museum, Hendon

Picture Credits

The author would like to thank the following for generously supplying a number of the photographs reproduced in this book: John Bell, Jan Cheney, Colin Cole, Charles Foster, Benny Goodman, Sydney Grimes, John Gumbley, Gerry Hobbs, Johnny Johnson, Klaus Rohwedder, John Saunders, Mary Stopes-Roe, Fred Sutherland, Frank Tilley and Rusty Waughman. Other photographs are from the author's private collection and the following sources:

INTEGRATED

Robert Owen, Official 617 Squadron Association Historian:
    pp. 1, 57, 102, 128, 203, 227, 293, 300, 309
National Archives, Kew: p. 160
Popperfoto/Getty Images: p. 233

PICTURE SECTION

Robert Owen, Official 617 Squadron Association Historian: p. 1
    (top), p. 3 (bottom), p. 4 (top), page 5 (bottom), p. 6 (top)
Anna Gowthorpe/PA Press/Press Association Images: p.1
    (bottom)

Index